Understanding
Medications

Understanding Medications

What the Label Doesn't Tell You

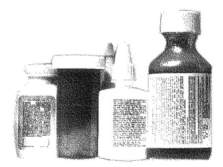

Alfred Burger
University of Virginia

American Chemical Society 1995

Library of Congress Cataloging-in-Publication Data

RM
301.15
.B873
1995

Burger, Alfred, 1905–

Understanding medications: what the label doesn't
   tell you / Alfred Burger.

      p.   cm.

      Includes index.

      ISBN 0–8412–3210–5 (hc: alk. paper).—ISBN
0–8412–3246–6 (pbk.: alk. paper)

      1. Drugs—popular works.   I. Title.

RM301.15.B873   1995
615'.1—dc20   95–17498
                                                        CIP

This book is printed on acid-free paper.

About the Author

ALFRED BURGER conceived and founded the *Journal of Medicinal Chemistry* in 1959 in collaboration with Arnold H. Beckett. From 1959 until his retirement in 1971, he served as editor of the journal.

Born in Vienna, Austria, he entered the Vienna Schottengymnasium, which taught a curriculum stressing the ancient languages but did not teach chemistry. He entered the University of Vienna as a student of chemistry and pharmacy in 1923. In 1928, he received his Ph.D. degree. In 1929, he went to work at the Drug Addiction Laboratory of the National Research Council, which had been established at the University of Virginia. In the Drug Addiction Laboratory, his chemistry was devoted to molecular modifications of model ring systems as they occur in the morphine alkaloids. He worked on the ring enlargement of cyclic ketones with diazomethane—a reaction that has become a classic in this field.

Burger married Frances Page Morrison in 1936 and established roots in Albemarle County. When the Drug Addiction Laboratory moved to Bethesda, Md., as the Laboratory of Chemistry of the National Institute of Arthritis and Metabolic Diseases in 1938, Dr. Burger remained at the University of Virginia as an assistant professor of chemistry. He advanced to associate professor in 1946, professor in 1952, and was chairman of the Department of Chemistry in 1962–1963. During this time, Burger's research was guided by pharmacological and therapeutic considerations of isosteric structures in drug design. His research led to almost 800 new compounds that were subsequently examined for their biological activity. One of these, the clinically useful antidepressant tranylcypromine, is currently marketed in the United States.

Burger has traveled widely as a result of site visits and lectures on behalf of the American Chemical Society and the American Pharmaceutical Association. He has earned many awards, including the Louis Pasteur Medal, the Distinguished Service Award of the Virginia Section of the American Chemical Society, the Smissman Award in Medicinal Chemistry, and the Award in Medicinal Chemistry of the American Pharmaceutical Association. He was general chairman of the Third Medicinal Chemistry Symposium in Charlottesville, Va., in 1952, chairman of the Division of Medicinal Chemistry of the American Chemical Society in 1954, and chairman of the Gordon Research Conference in Medicinal Chemistry in 1959. The American Chemical Society has named its biannual Award in Medicinal Chemistry for him.

His main hobby is music. He has played the violin since he was 11 and has been a member of the University of Virginia Orchestra and many other amateur orchestras and chamber music groups.

Contents

Preface

I live in a community of retired people who are better educated and more alert to innovations than the average men or women of any age. They carry on interesting conversations about art, economics, history, investment tactics, music, politics, and other topics that formed the basis of their liberal education in college long ago. As the only medicinal scientist among them, they expect me to express expert opinions on topics ranging from space flight and atomic fission and fusion to the meaning of various ailments that should not be serious enough to send them to a physician and encumber their Medicare accounts.

These people are curious about drugs. Does Tylenol cause addiction, and should they therefore take ibuprofen for relief from minor pain? Is ibuprofen just another name for aspirin? After all, these drugs are used for the same purpose, aren't they? And how does aspirin know where to go after one swallows it? One can take aspirin for a headache or for rheumatism in the leg; does the painful place call the drug to its site? What about vitamins? Is it a good idea to take a vitamin pill every day if I cannot stand eggplant or string beans? And what should one think of the TV report that fat in the diet promotes the occurrence of malignancies?

These and many other queries prompted me to write about drugs to satisfy and stimulate the curiosity of educated lay people who, on the whole, lack the biochemical and pharmacological background for the study of medicinal agents. I hope that this volume will provide some of the answers.

THE SCOPE OF MEDICINAL SCIENCE

The British Nobelist Sir Peter Medawar wrote, "We all nowadays give too much thought to the material blessings or evils that science has brought with it, and too little to its power to liberate us from the confinements of ignorance and superstition." Like scientists in other fields, medicinal scientists want to know about the mysteries that hamper the understanding of what appears to be or should be an ordered sequence of facts and observations.

This longing for a predictable order is based on faith that nature can be understood by applying reason. At every step toward this goal we have found that the more we know, the more we do not know. Beyond this quest for satisfying our curiosity about nature, the medicinal scientist searches for avenues to useful products and processes that could advance the field of medicine. Thus, medicinal chemistry, pharmacology, microbiology, and other subdivisions of medicinal science can become applied sciences which, in turn, can increase the basic understanding of natural phenomena.

Biologists, chemotherapists, and other medicinal researchers do not work and study behind closed doors like the scientist-monks of yesteryear. They are at their best when solving problems as cooperating teams in laboratories and research clinics that are open to discussion and the exchange of ideas. The educated layperson or journalist is received as a welcome partner when the solution of a problem has reached a stage that overlaps with the language of the public. The public furnishes the often expensive financial support for the scientists' work, and the public is eager to appreciate the advances that may ultimately be translated into new methodology in the treatment of diseases.

All drugs are chemicals, most of them organic chemicals, either natural or synthetic. Like all chemicals, they can decompose under the influence of heat, light, or moisture. Drugs can react with other chemicals either in a laboratory container such as a test tube, or in the surrounding of natural cells or tissues, whether of microbial, plant, animal, or human origin. The missions and the fate of chemical drugs are explained in this book to answer questions about why we use drugs.

The section on drugs of abuse, which have given the term *drug* a bad name, treats their origin and performance the same way as the sections on medicines for heart failure, asthma, or kidney ailments.

However, causes of drug abuse, both biological and sociological, are also considered.

Personal experience helps a physician in choosing a prescription for a given patient. If two patients present the same complaint and the same apparent symptoms, it is easy to prescribe the same drug in both cases. If these nuances differ and if the lifestyle and the diets of the two patients are not the same, the dosage or even the specific drug may have to be altered.

Too little attention is paid in medical education to drug interactions and drug incompatibilities. Until recently, the role of nutrition and diet in the control of disease has also not been taught adequately in the medical school curriculum. Because this oversight has been corrected in most medical schools, younger doctors may have better insight into these important problems.

Most effective medications are available to the patient by means of prescriptions from licensed physicians. The hundreds of over-the-counter (OTC) drugs lining the shelves of drugstores and grocery stores can be obtained without a prescription. Most of the OTC medications contain long-established chemicals with a minimum of side effects and are fairly, though never totally, safe.

This book is not a household compendium of medicines that can be consulted if one experiences a headache, stomachache, fever, or other symptoms of some underlying abnormal condition. Self-medication remains a hazardous procedure because the recommendations of the drug-packet insert are by necessity general and cannot anticipate every individual variation.

ACKNOWLEDGMENTS

Medicinal sciences have always fascinated lay people who want to understand the drugs they use and drugs that are abused. This volume is based on a small book entitled *Drugs and People: Medications, Their History and Origins, and the Way They Act,* published by the University Press of Virginia in 1986; a paperback minor revision followed in 1988.

The field of therapeutic agents and of drugs of abuse has expanded rapidly in the past decade. Many methods of study and lines of research that barely existed 10 years ago have altered medicinal science to such an extent that a radical reorganization and revision of the earlier books appeared necessary. I am grateful to the University Press of Virginia for enabling me to publish this new version under new sponsorship, to Barbara E. Pralle of the American Chemical Society Books Department for arranging the publication, and to Paula M. Bérard also of the ACS Books Department for her skillful and critical editing of the manuscript of this book.

ALFRED BURGER
January 15, 1995

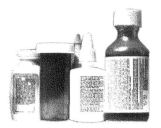

1

Drugs: Historical Beginnings

Drugs are chemical compounds that modify the way the body works. Most people think that these biological activities should help or heal sick people or animals. There is, however, no known drug that is not harmful or even poisonous at high doses, and much of the scientific work on drugs has attempted to widen the gap between effective and toxic doses.

The word *drug* has acquired bad connotations in recent years because the widespread abuse of a few chemicals that affect the central nervous system has become a serious sociological problem. Nevertheless, drugs act on many other organs in the body, can benefit as well as harm the nervous system, and have made possible a revolution in the way modern doctors treat disease.

Just as there is no health benefit without potential toxicity, there is no absolute goodness about drugs. However, their enormous health benefits outweigh the drawbacks in individual cases. The history, discovery, manufacture, action, acceptance, and rejection of drugs are the themes of this book.

It used to be said that what distinguishes humans from animals is that people take drugs. This old adage is no longer quite true. Rats and monkeys that have been addicted experimentally to some drugs will inject themselves with those drugs to support their addictions. But otherwise the old saying still holds.

The history of drugs is shrouded in the beginnings of the human race. Alcohol was made, drunk, and used to excess as far back as memory and records go. Tobacco (*Nicotiana*), hemp (*Cannabis sativa*), opium poppy (*Papaver somniferum*), and other plants containing drugs have been chewed and smoked almost as long as alcohol, and coffee has been served in the Middle East throughout that area's history.

Tobacco was carried from Virginia to England by Sir Walter Raleigh, whose pipe smoking prompted Elizabeth I to remark, "I don't like this herb." Of course, the queen did not know anything about tar and nicotine, but she became one of the first people to initiate the acrimonious debate about tobacco constituents that we face today. Likewise, the effects of cannabis have given it a bad name. Coffee was introduced by the Ottomans to the Western world when the Turks made a foray into central Europe in the 16th century. Its active alkaloid, caffeine, is often on the forbidden list for patients suffering from rapid heart beat or angina. Some of the chemicals that flavor coffee, such as esters of caffeic acid, stimulate cardiac hormones and thus add to the danger of disturbing the rhythm of the heartbeat.

Apes and humans are believed to have taken their separate evolutionary ways some 5–10 million years ago. In those times, prehumans, almost humans, and later, nomadic groups of obviously human individuals roamed the landscape in search of food and shelter. The driving forces behind this foraging behavior were the need to defend themselves against the environment and the need to reproduce their species. The earliest humans were often threatened by malnutrition or even starvation; by predators—both animal and competing or cannibalistic human hunters; and by parasites and degenerative diseases. The kindling and taming of fire further separated humans from animals. Fire gave primitive nomads some protection from cold and a way to make food more palatable.

The earliest records of their short and deprived years, estimated as no more than a couple of decades, are wall paintings and carvings on rocks that have endured for thousands of years in jungles or deserts. The pictures give us little insight into any healing arts those forebears may have invented. Probably accidental discovery of the healing powers of roots, barks, leaves, and berries and of nutritional sources of proteins and starch occurred even during those earliest stages.

Only recently, 10,000–20,000 years ago, nomadic tribes began to settle down in some parts of the world and formed small agricultural communities. This development was one of the most profound changes in the history of the human race. Planting seeds, domesticating animals, and erecting permanent shelters for humans and animals improved nutrition, provided more comfortable, though still primitive, living accommodations, and started the societal bonds that developed into villages, towns, and cities, where the first truly historical times began.

At this stage, when people began to live close together, certain rules were established to bring needed order out of unstructured conglomerated living. Strong individuals emerged as the leaders of tribal or village communities. Ambitious leaders increased their power by shrouding their personalities and decisions in mystery, elevating them to an exalted reputation. Thus emerged the concept of superhuman beings, of gods with powers over natural phenomena such as fertility, family bonds, lightning, and the awesome power of storms. When the demands of multiple deities for sacrifices grew excessive, all these powers and supernatural beliefs and superstitions were united in monotheistic religions, often after catastrophic wars and through a longing for peace. The deities were represented ritualistically by priests and priestesses, and valuable sacrifices offered by the people further enhanced the splendor and power of these leading personalities.

Part of the influence of these commanding individuals came from the personal help they could give to sick, wounded, and otherwise afflicted members of their groups. The medicine man, witch doctor, or shaman was the first person to turn to in distress. Sometimes authority lay in an influential woman who functioned as priestess, nurse, and midwife. In most cases, power, enforced by superstition, centered in a dominant male recognized as healer, priest, judge, and leader in peace and war. A knowledge of healing herbs formed part of his power base and prerogative; among his duties was applying such healing products when needed.

Botanical specimens taken from trees, shrubs, and other plants formed the mainstay of most of these early medicines. Sometimes the shaman deepened the mystery of the power of his medication by adding parts of animals, and even human hearts. Some mind-induced (psychosomatic) illnesses were probably improved by suggestions, just as they are today. Also, some functional disorders responded to the active chemicals in suitably chosen plants. It must have been hard for a medicine man to decide which plant to give a patient. Trial and error was the order of the day.

Even after a healing effect was found in a certain plant, it was probably used for a variety of illnesses, whether identical in symptoms or only vaguely similar. Modern physicians still face complicated choices in diagnosis and drug therapy in spite of help from analytical tests, X-rays, ultrasound, nuclear magnetic resonance visualization, blood cell counts, urinalysis, and other laboratory aids. We can only imagine the dilemma of an ancient medicine man, who was guided only by vague symptoms such as a generalized pain, nausea, fever, or convulsions. In such a dilemma, and without any knowledge of anatomy or pathology, some herbal concoction was given in the hope that it would work. The medicine man undoubtedly added prayers or exorcisms to the medication and believed sincerely that his ministrations would aid the afflicted.

Cynical healers also appeared. In Roman times, predictions about world affairs as well as the course of a patient's disease were based on the inspection of the intestines of newly slaughtered chickens. There was a saying that described one of these healer–seers (called a *haruspex*) encountering another member of the guild: *Haruspex, haruspicem videns ridet* ("When one haruspex sees another, he laughs").

EARLY RECORDS OF NATURAL DRUGS

Fortunately, some medicine men and women were careful observers, who had a patient's recovery uppermost in mind. Especially those who had risen to power and influence and had a scientific bent or deep compassion could be relied upon to search for valid explanations of their findings.

One of the oldest records of such medicinal recommendations is found in the writings of the Chinese scholar–emperor Shen Nung, who lived in 2735 B.C., or 4730 B.P. (before the present). He compiled a book about herbs, a forerunner of the medieval pharmacopoeias that listed all the then-known medications. He was able to judge the value of some Chinese herbs. For example, he found that *Ch'ang Shan* was helpful in treating fevers. Such fevers were, and still are, caused by malaria parasites. The drug consists of the powdered roots of a plant in the breakstone family (Saxifragaceae, now identified as *Dichroa febrifuga*, Lour.). Almost 4700 years later, a group of Chinese chemists isolated two compounds (the dichroines) from the plants, one of which later proved to control bird malaria. The leaves of this plant—called *Shun Chi* or *chuine* in present-day China—also contain antimalarial chemicals (the febrifugines), one of which is identical with one of the dichroines. These alkaloids (organic bases) were studied and synthesized during World War II in an effort to protect Americans from malaria in the Pacific and other tropical campaigns. However, chemists could not separate the nausea the drugs produced from their antimalarial effects.

The emperor Shen Nung also observed the stimulating effect of another ancient Chinese medicinal plant, *Ma Huang*. This one, now called *Ephedra sinica*, contains a number of alkaloids, chief of which, ephedrine, was isolated by the Japanese chemist Nagai in 1887, more than 4600 years after the effect of the plant had been recorded. Ephedrine and some closely related compounds are responsible for the ability of the wiry plant to stimulate blood pressure and breathing. The drug also contracts blood vessels. The Chinese book *Pen Tsao* (*Herbal Medicines*) of 5000 B.P. recommends castor oil for purging and the opium poppy for sleep.

The 5000-year tradition of natural drugs continues to flourish in China even today. A curious philosophy underlies their reasoning: Diseases are evil occurrences that are counteracted by good influences, they say, and nature is good and therefore healing. The traditional Chinese physicians overlook the fact that nature is not always kind. Cataclysms such as earthquakes, floods, and hurricanes threaten life. Medicinal plants are toxic, often deadly poisonous. The widespread toxicity of botanical materials can be seen in the history of assassinations, suicides, and accidental poisoning by swallowing certain plant products, as well as the killing of game and enemies with curare-tipped arrows.

Some Chinese pharmacognosists claim that synthetic compounds with no structural resemblance to harmless natural products are more dangerous than naturally occurring plant constituents, but this claim is unsubstantiated. Some synthetic compounds are highly toxic, others are not, and the same is true of natural products. In thousands of cases, chemical manipulation of their structure has lessened the toxicity of natural alkaloids, antibiotics, hormones, snake venoms, and other biologically active substances. Frequently, these semisynthetic chemical cousins or analogues can be used in clinical medicine much more safely than their natural ancestors.

Much knowledge of early drugs has been lost from every civilization. What remains is passed on in sporadically recorded epics and folklore unearthed by archaeologists and linguistic scholars. Tropical and subtropical regions, with their greater variety of plants, have given us most of the descriptions of these medicines. Although some ancient drugs have survived throughout the ages and are still used in a refined form, they amount to a small percentage of modern medications.

Ancient Hindu records mention eating chaulmoogra fruit to treat leprosy. We now know that the fruit contains several oils not very effective against leprosy bacteria. Treating the disfigured areas with these oils has been replaced entirely by swallowing dapsone, a synthetic drug, or by using other medicines.

A treatment still in use in underdeveloped regions for some intestinal problems is ipecac, the powdered roots of ipecacuanha (*Cephaelis* species). These plants were already being used by Brazilian Indians before the European conquest. The Indians treated "bloody flux," a form of amoebic diarrhea, with "igpecaya." Samuel Purchas (ca. 1577–1626) in *Purchas his Pilgrims* published a description of this material in 1615. As often happens in medicine, the successful recovery of a highly placed person made treatment with a new drug popular. In the case of ipecacuanha, the dauphin, son of Louis XV, was cured of his amoebic infection by the physician Claude-Adrien Helvétius (1715–1771). However, amebiasis was not studied intensively until 1875, and another 35 years passed before extraction of ipecac showed that the alkaloids emetine and cephaline were the active ingredients of the plant. Ipecac had been used for centuries to make poisoned people vomit; unfortunately, this property has limited the use of emetine for curing amebiasis.

Mediterranean peoples used to pulverize the dried flower heads of species in the ragweed family, especially *Artemesia maritima*, to get rid of intestinal worms. This plant is native to south-central Russia. Its active ingredient, santonin, was synthesized by British and Swiss organic chemists, yet studies have not been able to separate clinically effective doses from toxic ones. However, santonin is still used for roundworm infections in farm animals.

Another plant used to treat worms is goosefoot (*Chenopodium anthelminticum*). From its flowers comes a volatile oil that contains the active principle ascaridole. The Romans gave it the name Chenopodium; the Hebrews called it Jerusalem Oak; and others dub it Mexican tea.

Likewise, a thick, dark green oil that helps to expel tapeworms can be pressed from the male fern (*Dryopteris filix-mas*). This plant's use is ancient. Its active ingredient is also ascaridole. In humans it is effective only at almost toxic doses, and it is prescribed mostly in veterinary medicine. Early medicine men used it because they needed medications regardless of side effects. Many of these antique natural drugs would be unacceptable to today's regulatory agencies such as the U.S. Food and Drug Administration, as well as to the medical profession.

South and Central American Indians made many prehistoric discoveries of drug-bearing plants. Mexican Aztecs even recorded their properties in hieroglyphics on rocks, but our knowledge of their studies comes mainly from manuscripts of Spanish monks and medical men attached to the forces of the conquistador Hernán Cortés (1485–1547).

Pre-Columbian Mexicans used many substances, from tobacco to mind-expanding (hallucinogenic) plants, in their medicinal collections. The most fascinating of these substances are sacred mushrooms, used in religious ceremonies to induce altered states of mind, not just drunkenness. In the recent past, many of these mushrooms as well as flowers and shrubs have been extracted chemically, and their active ingredients have been identified. For example, peyote, a small cactus, now named *Lophophora williamsii*, contains alkaloids, especially mescaline, that cause hallucinations. The Indians called sacred mushrooms *teonacatl* (*nahuatl* means God's flesh). Some of these mushrooms belong to *Psilocybe* species and contain hallucinogens (psilocine, psilocybin). These and other plants that produced temporary psychotic reactions were abhorred by the Spanish priests, who saw their use as rites of the devil.

Other South American Indians, especially those in the Peruvian Andes mountains, made several early discoveries of drug-bearing plants. Two of these plants contain alkaloids of worldwide importance that have become modern drugs. They are cocaine and quinine.

Cocaine is extracted from leaves, especially from *Erythroxylum coca*, a bushy shrub native in South American countries at high altitudes, such as Bolivia, Peru, Ecuador, and Chile. Cocaine is the primary alkaloid in these leaves. Sigmund Freud, the Austrian psychoanalyst (1856–1939), treated many deeply disturbed cocaine addicts. In the course of his practice, he noted the numbing effect of the drug. He called this effect to the attention of the clinical pharmacologist Carl Köller, who introduced cocaine as a local anesthetic into surgical procedures.

Cocaine's potential for addiction was known and used with sinister intent by South American Indian chiefs hundreds of years ago. The chiefs maintained a messenger system along the spine of the Andes to control their thinly populated kingdoms, which stretched for thousands of miles along the mountains and were isolated from each other by the rugged terrain. The messengers had to run at high altitudes and needed stimulants for this exhausting task. Their wealthy employers provided the runners with coca leaves for this purpose and enslaved them further by paying them with more coca leaves, thus maintaining the addiction for which the poor runners were willing to continue their never-ending jobs. When coca leaves reached Europe with the Spanish conquistadores, they led to one of the first European waves of euphoric hallucinogenic drugs.

Quinine was isolated from the bark of the cinchona tree by the French chemists Joseph-Bienamé Caventou and Pierre-Joseph Pelletier in 1820, 200 years after the bark was introduced into Europe for the treatment of malaria. The Peruvian Indians had recognized for years the value of the quinquina tree for treating feverish patients. Some historians believe that malaria was imported to South America by the conquistadores and their African slaves. A persistent story exists about Doña Francisca Henriquez de Ribera, wife of Count Chinchón, the Spanish viceroy of Peru. She fell ill with malaria (the "tertians" variety with chills and fever that recur every third day) and was cured by an Indian healer who gave her the bark. In gratitude for the cure, the countess distributed the bark to other patients in Lima and thus alerted Spanish physicians to its clinical potential. The great Swedish botanist Linnaeus (Carl von Linné, 1707–1778) later called the tree cinchona in honor of Countess Chinchón, mispelling her name in the process. It is improbable that the countess persuaded Spanish doctors to use the bark because she died in Cartagena, Colombia, in 1641, while returning home. Because the antimalarial value of cinchona became more widely recognized while supplies of the bark fell short of demand, the cost of the powder was often matched by its weight in gold.

The cinchona tree grows wild in the sub-Andean jungles, and a number of European powers tried to transplant it to other tropical places. Peruvian officials realized what a gold mine these trees represented and strictly prohibited their export. A British attempt to smuggle some trees out of Peru failed, but two Dutch adventurers managed to get a few specimens across the border. The stolen trees were taken to Java and became the ancestors of later improved plantation trees that, before 1940, furnished 97% of the world's supply of quinine.

The inaccessibility of Java—and of Sri Lanka, where a few smaller plantations existed—became a source of worry for European drug factories that were the principal sources of pure quinine. This concern was felt acutely by German manufacturers in World War I, when they were unable to supply European colonies in Africa with the drug. In World War II, British and American suppliers also were cut off from their Oriental sources of quinine when the Japanese occupied Java and Malay. In both instances, the drug shortage stimulated intensive research to surmount this handicap, and the resulting new compounds are almost the only effective synthetic antimalarials we have today. Nevertheless, quinine has kept a modest but important and inexpensive place in antimalarial treatment.

INORGANIC DRUGS

As the centuries unrolled and new civilizations appeared, cultural, artistic, and medical developments shifted toward the new centers of power. A reversal of the traditional search for botanical drugs occurred in Greece in the fourth century B.C. (about 2400 B.P.), when Hippocrates (estimated dates, 460–377 B.C.), the "Father of Medicine," became interested in inorganic salts as medications.

Hippocrates' authority lasted throughout the Middle Ages and reminded alchemists and medical experimenters of the potential of inorganic drugs. In fact, a distant descendant of Hippocrates' prescriptions was the use of antimony salts in elixirs (alcoholic solutions) advocated by Basilius Valentius in the middle of the 15th century and by the medical alchemist Phillippus Aureolus Paracelsus (born Theophrastus Bombast von Hohenheim, in Switzerland, 1493–1541). The ethics of Hippocrates as incorporated in the physicians' Hippocratic oath have survived better than his preference for inorganic salts.

However, we still use magnesium sulfate (named Epsom salts for the British town of Epsom), both internally and externally; aluminum salt astringents; sodium and potassium chlorides and calcium salts for various deficiencies; barium sulfate as an X-ray contrast agent; and sodium iodide to prevent thyroid disorders, as well as stannous fluoride to prevent tooth decay. Gold compounds are experiencing a renaissance in the treatment of arthritis, and silver nitrate was used to protect the eyes of newborn infants from gonorrheal blindness before penicillin took its place. Lithium salts are used in gout and to smooth out the biphasic phase of manic-depression. Many heavy metals are incorporated as traces in diet supplements because they have been recognized as essential parts of important biological catalysts.

The next great—and reactionary—influence on medicinal thought came from a Greek physician from Pergamum, Claudius Galenus, or Galen (129–ca. 199 A.D.), who taught in Rome. Galenic medicine consisted of preparations of plants by soaking (infusion) or boiling (decoction). As Oliver Wendell Holmes said, "These Galenists were what we should call herb doctors today." Galen claimed that herbal mixtures could provide all the essentials for health and therefore could be applied to all conceivable health defects. He was also a vegetarian.

We know today that a strict vegetarian diet, without milk, cheese, or eggs cannot contain all the protein-building amino acids needed for normal growth and body maintenance. In Galen's storeroom (*apotheke*), some metallic substances, such as copper and zinc ores, iron sulfate, and cadmium oxide, were still present, probably as a tribute to Hippocrates' drug inventory. Galen insisted on carefully identifying the type and age of botanical materials and thus foreshadowed the value of controlling the purity of drugs. Among his favorite and potent drugs were hyoscyamus (which contains atropine), opium (the source of morphine), and squill (which contains heart stimulants, cardiac glycosides similar to digitalis).

SOME MEDIEVAL MEDICINES

Galen's teachings made such a profound impact on medieval society and medicine that they were followed for more than 1000 years. In western Europe, where medical knowledge was encased in Catholic monasteries, Galen's prescriptions were embraced by the conservative monks. Although some teachings of contemporary Islamic scholars filtered into this environment, many centuries passed before herbal medicine could be replaced by newer treatments.

After the fall of the Roman and Byzantine empires and the rise of Islam in the Middle East, the "civilized world" of Europe was in decline. The migration of Eastern peoples, the constant and cruel wars all over the continent, and the absence of cultural and intellectual life-styles made scholarly studies of drugs impossible. The herb gardens of monasteries were the chief source of healing plants, but infections, heart disease, cancer, and the innumerable battle wounds and disfigurements from torture could not be treated rationally. Plagues swept across the continent and the British isles, chiefly bubonic plague, viral pandemics, and later syphilis. The victims were laid to rest in mass graves with no medicines to ease their final agonies. Life was hard and, on the average, short; at 50 or 55 a person was an ancient senior citizen; most died in their 40s or earlier.

Bleeding the patient by opening a vein or with leeches was one of the few medical treatments and was used for various ills. Amputation of wounded or badly infected limbs was carried out without benefit of anesthesia or soap-and-water hygiene; cesarean deliveries meant the cruel and certain death of the mother.

These desperate health conditions applied to kings and serfs alike for almost 1500 years. It is no wonder that a better life could be imagined only in the heavens, which represented a hypothesis that had never been proved.

The little that was known about healing plants, minerals, and tissues was called *materia medica,* a term still used for drug information at the turn of this century. Latin was used throughout the collective accounts of this subject because it was the professional language of the monks and also because it kept the common people in ignorance.

When pamphlets could no longer hold the accumulating knowledge of *materia medica,* larger, more formal collections were gathered in national pharmacopoeias. The first of these books appeared in Florence in 1498, six years after Christopher Columbus landed in Dominica, followed by others in Nuremberg (1535), Basel (1561), Augsburg (1564), and London (1618). Standards of purity and methods of preparing various drug products accompanied the descriptions of botanical and mineralogical specimens. One item tells how to make a sort of candy of red rose petals for pale tired people and of white roses for those with too ruddy complexions.

The late Middle Ages coincided with the upsurge of alchemy, a primitive chemistry dealing mostly with inorganic substances. The renewed interest in inorganic materials pushed botanical sources of medicines into second place temporarily. True, quinine (cinchona) appeared during that century, but the next great botanical drug, digitalis, was not introduced until the end of the 1700s.

2

Early Modern Medicines

DIGITALIS

In 1875 the British physician William Withering published a book entitled *An Account of the Foxglove and Some of its Medical uses: with Practical Remarks on Dropsy and Other Diseases.* Dropsy, characterized by the accumulation of fluid (edema) as with congestive heart failure, had previously been treated with dried and powdered purple foxglove (*Digitalis purpurea*), which is still grown as an ornamental flower. But digitalis had remained a typical Galenic drug used also for skin ulcers, epilepsy, and other poorly diagnosed diseases. Withering himself had no clear picture of the cause of edema but noticed in passing that digitalis powerfully affected the heart. It took another 14 years for John Ferriar to sort out the ability of digitalis to increase the contraction of heart muscles and to understand that its diuretic effect on the kidney, which increases the secretion and flow of urine, was secondary to this action.

The compounds responsible for the action of digitalis have complicated chemical structures. They are called cardiac glycosides, and they are found in a number of plants. They occur, for example, in squill, the dried, fleshy bulb of the sea onion (*Urginea maritima*), which was used as a remedy in ancient Egypt and as a heart tonic and kidney stimulant by the Romans. In stronger, more toxic doses, this material causes vomiting and has been used as a rat poison. In 1776, a Javanese king executed 13 unfaithful concubines. He had an executioner make an incision with an awl-like device, then apply sap of the upas tree, which contains cardiac glycosides, to the incision. The women died quickly in utmost agony.

The seeds of *Strophanthus* also contain glycosides and were used in African arrow poisons. Other sources of these heart stimulants are skins of common toads, which figured in ancient Chinese folk medicine and in European medieval prescriptions until they were replaced by digitalis. The purple foxglove had been known botanically since 1250, when it was mentioned by Welsh physicians, but it was named in 1542 by the botanist Fuchsius.

The individual digitalis glycosides are hard to separate, but a few have been purified. Fairly well purified preparations are available commercially, but because they cannot be reliably analyzed chemically, they must be evaluated biologically. For example, their effect on heart contraction is measured on the toad heart. A purified glycoside from another kind of digitalis (*D. lanata*), called digoxin and taken orally, is probably the most widely prescribed and satisfactory digitalis product in modern heart treatment.

Although the isolation of pure cardiac glycosides had to await the development of modern ways to separate closely related chemicals, the introduction of digitalis was one of the earliest steps in modern drug therapy. During the period following the American, the French, and the beginning of the Industrial Revolutions, shackles of medieval thought patterns were shed, and explanations of drug actions were sought. Several botanical drugs were therefore extracted, purified, and concentrated. Then the residue was chemically fractionated to isolate individual components of the usually gummy mixtures. The resulting fractions were used as drugs.

OPIATES

The first examples of natural products that fit into this account historically are opium and its alkaloids. Alkaloids are natural substances that are chemically alkaline or basic. Opium is obtained by cutting surfaces of unripe seed capsules of the oriental poppy (*Papaver somniferum*). A milky juice oozes from the wounded seed box and is dried in the air. The brown gummy residue is powdered. This flower spread from Asia

Minor to many Mediterranean and Southeast Asian countries. It is also grown in U.S. gardens for its brilliant bloom.

The word *opium* is derived from a Greek word meaning "juice." Even before the ancient Greeks encountered it, the Sumerians (Babylonians) carved tablets with pictures of the poppy about 6000 B.P., along with the inscription *hul* ("joy") and *gil* ("plant"), and apparently were aware of the mind-altering effects of opium.

The Greek naturalist Theophrastus (circa 372–circa 287 B.C.) mentioned poppy juice in his writing. During the rise of Arabian–Islamic medicine, opium was used widely and was introduced to Oriental peoples, principally as a constipant to control dysentery (much as we use paregoric). With Islamic inroads into Europe during the 16th century, opium was introduced there and became widely accepted. Paracelsus (1493–1544) made a purified preparation of poppy juice that he called laudanum, which can still be purchased in some drugstores today.

About 25 opium alkaloids are responsible for the drug actions of opium. The main one is morphine. It is a powerful *analgesic* (painkiller); in fact, it serves as a standard for all other natural and synthetic analgesics. Morphine also constipates by slowing down peristalsis, the involuntary muscle contractions that move the contents of the intestines on their course. It also depresses respiration, causes sleepiness, and has several other minor activities. Mentally, morphine produces euphoria, a sensation of well-being, and after repeated use, dependence liability (addiction).

A minor alkaloid of opium is codeine, which depresses the cough reflex much better than morphine. Codeine is still used in cough medicines, although it is usually replaced by synthetic compounds with fewer side effects. Codeine is an effective constipant but only a fairly good analgesic agent. It is slightly addictive.

Papaverine, another minor opium alkaloid, does not relieve pain, but because it dilates arteries well, it has some value in painful angina pectoris. Like codeine, papaverine is manufactured synthetically. Noscapine (narcotine) is a minor opium derivative that is beginning to enjoy acceptance as a cough medicine (*antitussive*). None of the other opium alkaloids are used as drugs, and many of them are toxic. Obviously, the various alkaloids in opium possess complex and often contradictory activities.

Some derivatives of morphine are stronger pain relievers than morphine itself, e.g., the synthetic drug heroin. Unfortunately, heroin is also more addictive than morphine and has become the most dreaded and abused pleasure-producing (euphoriant) drug. Heroin is made in one simple step by heating morphine with acetic anhydride, a common commercial chemical. Its introduction into medicine followed that of aspirin, which had been prepared by the German chemist, Friedrich Hofmann, in 1875 by heating salicylic acid with acetic anhydride (aspirin is acetylsalicylic acid). The German pharmacologist H. Dreser found in 1899 that aspirin was a better, less toxic analgesic than its parent compound, salicylic acid. It then occurred to him that a similar process should improve the much more potent analgesic, morphine. This chemical synthesis produced heroin, which is indeed one and a half times more potent in relieving severe pain than morphine and was used for several years in postoperative analgesia until its addictive potential was realized. Heroin has been outlawed by all modern nations.

Smoking opium in specially constructed pipes has been practiced for centuries in oriental opium dens and in French literary salons. A major wave of abuse of opium and its relatives flooded the world in the 20th century and shows no signs of receding. However, other chemicals such as cocaine and some synthetics with more psychosis-producing (*psychotomimetic*) properties have displaced opiates in some parts of the world. Of these, cocaine has become the most widely abused drug of our time. It is easily extracted from the leaves of the South American coca bush, in which it occurs in the form of salts of plant acids. The base is liberated with sodium carbonate or hydroxide and extracted with organic solvents. The crude white crystalline base is sold illegally as "crack," which is smoked by drug users. The hydrochloride salt of cocaine is aspirated by being placed in the nostrils or injected to produce hallucinogenic effects.

SIDE EFFECTS AND STANDARDIZATION OF DRUGS

No drug has just one effect. Medically desirable actions are invariably accompanied by side effects. Such near-toxic or clearly poisonous results must be balanced against the clinical advantages of substances. This phenomenon becomes apparent in animal experiments, although the final value of a drug is determined through clinical trials and broad experience with patients.

The progress of almost all new drugs is quite similar. The drug is announced with fanfare in medical journals, at scientific meetings, and by the news media. Although couched today in increasingly cautious language, claims of the drug's performance are usually encouraging. Within a year or two, wide use of the substance by millions of patients inevitably discloses unwanted side effects, some serious, and physicians and patients alike often become cautious of its use. Generally, the number of serious side effects is relatively small, and when put into proper perspective, the ratio of benefit to risk becomes acceptable. Then the drug wins a permanent niche in the arsenal of medicinal agents.

In many parts of the world, drugs are dispensed in apothecary stores by druggists behind counters. They sell chiefly medical supplies. Licensed pharmacists used to spend part of their time compounding drugs and additives as prescribed by physicians. In the United States, such stores are called drugstores, but they are really department stores, the successors to the general stores of old. Nowadays pharmacists buy drugs in ready-to-take forms from wholesale distributors or directly from pharmaceutical manufacturers, whose sales personnel (detail men and women) visit the drugstores and take their orders. The same detail people also visit physicians and leave samples and descriptive pamphlets about their company's drugs. Like ready-to-eat foods, most drugs have been standardized in this manner, and the labor of measuring and mixing is done by machines in the manufacturers' plants.

In the Western world, this standardization has existed for a long time. In the East, natural drugs are still largely preferred to synthetics. Perhaps tradition and a deference to the economics of horticulture play roles in this attitude. There have been similar cases of agricultural concern in Europe and the United States. One such case followed the introduction of synthetic indigo, a blue dye originally obtained by fermenting *Indigofera tinctoria*. This plant had been grown in highly profitable plantations in South Carolina that were ruined by the appearance of the purer tint of the cheaper synthetic indigo. Fortunately, other economic specialty crops soon took the place of the plant.

Tobacco planters belong in this category. Tobacco is a profitable crop and the basis of a vast agricultural and industrial enterprise. Several areas in the United States, Asia Minor, and Africa derive their principal revenues from tobacco (*Nicotiana tabacum*). After grain alcohol, tobacco is probably the most widely used and abused crude drug in the world, yet many of its constituents, especially the nicotiana alkaloids and tobacco smoke have a variety of toxic effects. The two-phase action of nicotine and the minor tobacco alkaloids consists of briefly stimulating and then depressing certain centers (*ganglia*) in the cerebral cortex of the brain. This action leads to an overall calming not unlike that produced by tranquilizers, a pleasant feeling desired by a large percentage of the world's population. Therefore, physical dependence on tobacco is almost as hard to break as dependence on alcohol, and the raw materials for both substances appear destined to be grown for profit interminably.

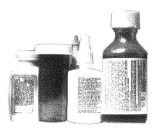

3

Naming Drugs

HOW DRUGS ARE NAMED

The unfamiliarity of drug names is a problem for many people; the names are difficult both to pronounce and to remember. Like the drugs themselves, their names have a history. The inorganic salts were named for the elements that form them. For instance, table salt is composed of sodium and chlorine; it is called sodium chloride. However, organic chemicals are named according to strict rules that have evolved over the past 150 years. In the beginning, medicines were named according to their sources. For example, atropine was named for its parent plant, *Atropa belladonna*; papaverine, for the poppy, *Papaver somniferum*; and adrenaline, for the adrenal gland. An early antifever (*antipyretic*) drug was called antifebrin; an antiseptic for the urinary tract, Urotropin. These names were coined by commercial manufacturers of the drugs and bore no relation to their chemistry.

Chemical names of organic molecules can be cumbersome. For example, the correct name of the simple structure of the antihistamine diphenhydramine hydrochloride, or Benadryl, is 2-diphenylmethoxy-1-*N,N*-dimethylaminoethane hydrochloride. Such names are systematized and brought up to date in *Chemical Abstracts*, a journal published by the American Chemical Society in Columbus, Ohio, and used worldwide. Who but a trained organic chemist could remember, pronounce, and decipher such chemical names? Not even pharmacists who dispense medicines, or physicians who prescribe them, or biologists who study them but have forgotten the short course in organic chemistry that they took in their sophomore year in college. For this reason, a new compound, while it is under biological study, is usually given a number and the initials of the chemist who prepared it or of the company where it was made. Thus, a compound might bear such code numbers as SK&F 385 (for SmithKline & French) or Win 8620 (for Sterling-Winthrop).

By the time a promising compound becomes a candidate for clinical trials, it often receives a name that is easier to use. Such names must be approved by the Council on Pharmacy and Chemistry of the American Medical Association; a few of them may in time become nonproprietary or generic names for the compounds. (See the table at the end of this chapter.) Most other developed nations approve similar generic names, and the World Health Organization in Geneva, Switzerland, publishes international nonproprietary names for drugs. The American Medical Association's Council adopts the generic name (which is called the U.S. adopted name, or USAN), then publishes it in the journal *New and Nonofficial Remedies*. These nonproprietary names may be used by any manufacturer who has the right to market the product; the names must appear on the package insert (the paper that is tucked in the box with a medicine), together with the dosage directions and other vital information, especially precautions and warnings of side effects. The public would do well to read these inserts.

The preferred way of forming the name of a drug is by a pleasant-sounding contraction of its chemical name. For example, the generic name of tranylcypromine (a drug developed to help depression, synthesized at the University of Virginia) is derived from its chemical name *trans*-2-phenylcyclopropylamine. The trade name of this drug is Parnate. The *-ate* ending comes from sulfate, the commercial marketing salt of the compound. Because the acid or the basic ion that forms the salt of an acid or a basic drug, respectively, is irrelevant for the action of the compound, its name is often dropped. The local anesthetic procaine is marketed as its hydrochloride salt, but this fact is seldom mentioned; the protected trade name of the drug is Novocain, but almost no physician or dentist will tell you that they will numb your tissues with Novocain hydrochloride.

Today in the United States, all rights to market a new drug may be protected for 22 years after the issue of a patent. During this period, the originating firm or inventor has exclusive rights to produce the drug. Quite commonly, the manufacturer guards his or her interest further by registering the proprietary name as a protected trademark. Such trade names may not be used by any other firm or person in marketing a drug even after the patent has expired. They are usually capitalized and followed by the symbol ®; the R stands for registered. Trade names are chosen to be pleasant sounding, easy to remember, and easy to pronounce.

Pharmaceutical manufacturers bombard physicians with advertisements, pamphlets, and especially visits from their sales forces, called *detail* men and women, who praise the advantages and uses of their proprietary drugs. Physicians, hard-pressed for time, frequently learn many of the essential facts about new drugs from these detail men and women, who naturally use the registered trade names in their descriptions. Usually the physician learns and remembers the trademarked name and hardly ever knows the more complicated generic name. Most prescriptions are written with proprietary names.

The proprietary names so widely used should be capitalized. But after several decades the original source of a drug sometimes becomes hazy and may be unofficially forgotten. Thus, Adrenalin, Aspirin, Atebrine, Novocain, Xylocaine, and other classical drug names were, and still are, proprietary trade names but are seldom capitalized in the current literature and in newspapers. After a patent has expired, any manufacturer may develop and market the drug, provided the manufacturer uses the less familiar generic name, and so long as he or she does not infringe on protected dosage forms, shapes, and colors. *In almost all cases, a generic drug is equivalent to the proprietary material; chemically the two substances are identical.* The U.S. Food and Drug Administration makes its own analytical tests and rejects inferior products.

But judging the availability of the drug in the body is difficult. A chemical is usually encased in a capsule or pressed in a tablet, often with inert diluting fillers, and therefore may dissolve at a different rate in body liquids than does the original proprietary product. This potential defect, although rare, may affect the availability of the drug to body tissues. Again, our regulatory governmental agencies are charged with keeping a wary eye on this possibility by measuring the speed of solution and the effects of supposedly inert additives.

Physicians can consult the annual editions of the *Physicians Desk Reference (1)* for up-to-date drug names and other pertinent information. Names are listed in both generic and proprietary indices; often a drug developed simultaneously by different firms in different parts of the world may have different names. A more unified naming system is badly needed.

The member countries of the World Health Organization have been asked to observe certain recommendations. As with other recommendations, they are not always heeded, but they are an attempt at international uniformity.

1.   Names should preferably be free from any anatomical, physiological, pathological, or therapeutic suggestion.
2.   Names should be formed by combining syllables from the chemical name of the compound in such a way that its important chemical groups are indicated.
3.   In general, names should be distinctive in sound and spelling, should not exceed four syllables, and should not easily be confused with names already in use. They should not end with a capital letter or a number.

SOME ALTERNATIVE DRUG NAMES

Depending on their origin, history, and manufacture by competing companies, drugs often have alternative proprietary or common names. The names in the first column are nonproprietary; most of those in the second column are proprietary names, chosen at random and without prejudice. Trade names appear in upper-case letters. In many cases they represent the major commercial names in the United States.

Some Nonproprietary Names	Selected Alternative Names
acetaminophen	Tylenol
acetylsalicylic acid	aspirin
acrisorcin	Akrinol
amantadine	Symmetrel
aminacrine	aminoacridine
aminocaproic acid	Amicar
amobarbital	Amytal
amphetamine	Benzedrine
amylocaine	Stovaine
ascorbic acid	vitamin C

Some Nonproprietary Names	Selected Alternative Names
bacitracin	Neosporin
barbital	Veronal
benzalkonium	Zephiran
bialamicol	Camoform
carbarsone	amebarsone
cetylpyridinium chloride	Cepacol
chloramphenicol	Chloromycetin
chlordantoin	Sporostacin
chlordiazepoxide hydrochloride	Librium
chloroguanide hydrochloride	Proguanil
chlorophenothane	DDT
chloroquine	Aralen
chlorothiazide	Diuril
chlorpromazine	Thorazine
chlortetracycline hydrochloride	Aureomycin
cimetidine	Tagamet
cyanocobalamin	vitamin B_{12}
cycloguanil	chlorguanide triazine
cyclophosphamide	Cytoxan
dapsone	Diphenasone
desipramine hydrochloride	Norpramin
dexamethasone	Decadron
dextroamphetamine sulfate	Dexedrine
diazepam	Valium
dichlorophen	Parabis
dicumarol	Dicoumarin
diethylcarbamazine	Hetrazan
diethylstilbestrol	DES
digoxin	Lanoxin
dimenhydrinate	Dramamine
diphenadione	Dipaxin
diphenhydramine hydrochloride	Benadryl
diphenoxylate hydrochloride	Lomotil
doxorubicin hydrochloride	Adriamycin
epinephrine	Adrenalin

Some Nonproprietary Names	Selected Alternative Names
erythromycin	Erythrogran
estradiol	Progynon
ethacrynic acid	Edecrin
ethionamide	Trecator-SC
famotidine	Pepcid
felbamate	Felbatol
finasteride	Proscar
fluotexine	Prozac
flurazepam hydrochloride	Dalmane
furosemide	Lasix
glaucarubin	Glaumeba
griseofulvin	Fulvicin
halazone	Pantocid
haloperidol	Haldol
heparin sodium	Hep-Lock
hexachlorophene	pHisoHex
hydrochlorthiazide	HydroDIURIL
hydroxytryptamine	5-HT, serotonin
hydroxyurea	Hydrea
ibuprofen	Motrin, Advil
imipramine hydrochloride	Tofranil
indomethacin	Indocin
iodochlorhydroxyquin	Entero-Vioform
iproniazid	Marsilid
isocarboxazid	Marplan
isoproterenol	Isuprel
kanamycin sulfate	Kantrex
kaolin and pectin	Kaopectate
leucomycin	Kitasamycin
lidocaine	Xylocaine
lincomycin hydrochloride	Lincocin
loperamide hydrochloride	Imodium A-D
lucanthone hydrochloride	Miracil D
lysergide	LSD
melarsoprol	Mel B

Some Nonproprietary Names	Selected Alternative Names
meperidine hydrochloride	Demerol
mephobarbital	Mebaral
meprobamate	Equanil, Miltown
meralluride	Mercuhydrin
mercaptopurine	Purinethol
methadone	Dolophine
methamphetamine	Methedrine
methotrexate	Folex
methyldopa	Aldomet
metronidazole	Flagyl
nialamide	Niamid
niclosamide	Niclocide
norepinephrine bitartrate	noradrenaline
novobiocin	Albamycin
oleandomycin	Amimycin
oxytetracycline	Terramycin
pargyline hydrochloride	Eutonyl
phenelzine sulfate	Nardil
phenobarbital	Luminal
phenoxymethylpenicillin	penicillin V
phenytoin	Dilantin
prednisolone	Paracortol
prednisone	Paracort
primidone	Mysoline
procainamide hydrochloride	Pronestyl
procaine hydrochloride	Novocain
progesterone	progestin
propranolol hydrochloride	Inderal
pteroylglutamic acid	folic acid
pyrazinamide	Aldinamide
pyridoxine hydrochloride	vitamin B_6
pyrimethamine	Daraprim
quinacrine	Atebrin
reserpine	Serpasil
retinol	vitamin A

Some Nonproprietary Names	Selected Alternative Names
riboflavin	vitamin B_2
rifampin	Rifadin
secobarbital	Seconal
spiramycin	Sequamycin
stibocaptate	Astiban
stibophen	Fuadin
sulfamethoxazole	Gantanol
sulfisoxazole	Gantrisin
sulindac	Clinoril
suramin sodium	Antrypol
terazosin hydrochloride	Hytrin
tetracycline hydrochloride	Achromycin
theophylline	Respbid
thiamine hydrochloride	vitamin B_1
thiopental sodium	Pentothal
thioridazine	Mellaril-S
tocopherols	vitamin E
tolnaftate	Tinactin
tranylcypromine sulfate	Parnate
triamterene	Dyrenium
trimethoprim	Proloprim
tryparsamide	Tryparsone
tylosin	Tylan
undecylenic acid	component of Desenex
valproic acid sodium salt	Depakene
verapamil	iproveratril
vitamins K_2	menaquinones
warfarin sodium	Coumadin

REFERENCE

1. *Physicians Desk Reference;* Medical Economics Data: Montvale, NJ, published annually.

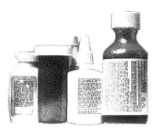

4

Biomedical Research

Progress in science seldom occurs in giant steps. Rather, one small step forward is discovered here, another one there. Once in a while an innovative young scientist puts those scattered findings together, sees how they fit, and from them makes predictions. In the case of biologically active chemicals, the age-old question of how they really work has not yet been answered, but many medicinal scientists have offered opinions, based on more or less good, experimental evidence. Take the case of time-honored aspirin. The chemical name of aspirin, acetylsalicylic acid, comes from the Latin botanical name for the willow (*Salix alba*). For hundreds of years a sort of willow tea relieved aches and pains. But it took 71 years from the introduction of aspirin as a drug in 1899 to reach the first inkling of an explanation in 1970 of how its antiinflammatory action worked. Indeed, this information had to wait for a totally unrelated event, namely, the discovery of the way in which cells biosynthesize an inflammatory, hormonelike chemical, one of the prostaglandins. This example shows the truth that apparently abstract biochemical research may pay off in practical results.

A person who hunts with a gun or a bow and arrow aims at a target and tries to hit it. Until a few years ago, and only in a few cases since then, no one could recognize the target of a drug, and the aim therefore remained uncertain. There had to be a target, and it had to be a chemical. All drugs are chemicals, and they can only decompose or react with other chemicals in the cells of an animal. The drug has to be recognized and received by the cells; that location is called a receptor.

The earliest attempt to suggest the nature of a receptor took place at the turn of the century. It can be credited to Paul Ehrlich (1854–1915), the father of medicinal science. However, it took 95 years before receptors could be isolated in minute quantities and cloned to make enough receptors available for research. Most of the cell receptors turned out to be complex proteins, a few others are *nucleotides* (building blocks of nucleic acids). The chemical structure of a few receptors has been established by X-ray crystallography.

When these structures are projected on a computer screen, it becomes possible to construct simpler, traditional organic formulas that contain a few analogous functions at comparable points. The structures can be superimposed on the structure of the receptor on the computer screen and may, in some instances, be tested as blocking agents of the receptors. If the receptor belongs to the membrane of a nerve cell, a cancer cell, or an infectious bacterium or virus, the superimposable chemical may block the biological function of the receptor and become a healing agent in a disease caused by the cell.

In a similar manner, enzymes in the membrane or interior of a parasitic disease-causing cell may be prevented form sustaining cellular life. Again, drugs mimicking essential functions of the enzyme may interrupt the life of a cell that depends on the enzyme.

How do enzymes work? They greatly speed up chemical reactions; that is, they are *catalysts*. To appreciate their role in metabolism, we can think of a meal we eat and how it is *oxidized* (burnt up) in the body to furnish heat, that is, energy. If we imagine burning up the same meal on a burner, how much heat and how much time would that take? Depending on the size of the burner, complete burning in air (*combustion*) may take a long while. The same oxidation reaction, catalyzed by enzymes, proceeds at human body temperature (37 °C) and is completed in a few seconds. The energy created by this rapid oxidation is used to keep the body warm and to allow people to walk, work, and live it up.

In a reaction over a burner or open fire, the components of the material must be heated to high temperatures before they undergo a reaction with the oxygen of the air. How do enzymes perform the same task much more rapidly and under much milder conditions? In such catalyzed reactions the component chemicals (such as proteins and carbohydrates) fit themselves snugly to the active site of an enzyme molecule, so snugly that only a minimum of energy is required to initiate an interaction. When the materials of a meal have been oxidized in this way, the enzyme molecule retreats from close contact with those materials and is ready to oxidize another batch of the molecules.

Some enzymes can reduce a substrate (enrich its hydrogen content), some can join (*condense*) two substrates, and some can ferment and decompose a substrate. In every case, close contact of substrate and enzyme is needed to facilitate rapid interaction under mild conditions. Some enzymic reactions proceed up to 40,000 times faster than the same reaction without catalysis.

At the turn of the 20th century, hypotheses about the mode of action of drugs were almost nonexistent. Since Louis Pasteur's days (1822–1895), microbial diseases had been of prime interest to medical investigators. However, scientists needed to see microbes to follow their life processes with some certainty. A number of early microbiologists succeeded in selectively staining microbes without equally staining the surrounding tissues. The most durable of these staining methods was published in 1884 by Hans C. J. Gram (1853–1938). Paul Ehrlich and others used dyes to visualize microbes under the microscope. It occurred to Ehrlich that if such dyes were synthetically linked with toxic atoms or groups of atoms, they might kill dangerous microbes with minimal damage to the tissues of the host. This concept became the basis of *chemotherapy*, a term Ehrlich coined in 1891. Even though using the selective toxicity of dyestuffs in therapy is largely a thing of the past, drug selectivity has remained the prime goal of all medicinal science.

THE ROLE OF THE PHARMACEUTICAL INDUSTRY IN DRUG DEVELOPMENT

The first concept in developing a new drug arises in the mind of one clever inventor. But from there on, many minds and hands become involved. The compound must be synthesized or isolated by chemists in sufficient quantity and as inexpensively as possible. Relations of chemical structure to biological activity must be explored by making chemical analogs, often hundreds of them, to identify the most power-

ful, least toxic, and most suitable member of a series of related chemicals. The chemists work in close cooperation with pharmacologists and
microbiologists, who must develop fast, practical, and valid tests for
screening all those compounds.

This kind of cooperation is hard to find in university settings.
Intellectual freedom as it is found in academic research departments
often lures scientists into personal side roads. The chemists may want
to study a variety of interesting reactions of their compounds; the
biologists, located in other parts of town or even in other cities, may
undertake the study of cellular and microbial matters unrelated to the
development of the drug. These excursions delay and sidetrack the
main objective of the cooperative work areas, although they might
lead to valuable insights in other fields.

The only two types of institutions in which close cooperation
takes place between scientists in different areas of medicinal research
are private and governmental research institutes and the pharmaceutical industry. Private foundations provided early sources of support
for research in which organic chemists, toxicologists, microbiologists,
pharmacologists, biochemists, parasitologists, and physicians cooper-
ated. In such close quarters, ideas are constantly exchanged, with the
result that chemists become at least token biologists; physicians grope
for the underlying principles of their art; and experimental biologists
begin to think in biochemical terms. Yet an atmosphere of academic
freedom in such institutes allows and even encourages scientists to
explore questions that might have a bearing on their common goals.

Institutes of this type have sprung up in many parts of the
world, wherever the wealth of private individuals, foundations, or
governmental support has made them possible. The Pasteur Institutes
in Paris and other locations, the Salk Institute in La Jolla, California,
the National Institutes of Health outside of Washington, D.C., and the
Central Drug Research Institute in Lucknow, India, are some examples. Laboratories with a more narrowly defined goal but equally high
scientific standards include the Sloan-Kettering Institute for Cancer
Research in New York, the Southern Research Institute in Birming-
ham, Alabama, the Fox Chase Cancer Center in Philadelphia, Pennsyl-
vania, the Institutes for Tropical Diseases and Cancer in London, some
of the German Max Planck Institutes, and the Istituto Superiore de
Sanità in Rome. Some of these institutes have links with large hospitals but not at the expense of their research. Rather, they have become

a breeding ground for outstanding medical discoveries and have harbored several Nobel Prize winners.

Intermediate between these institutes and the strictly goal-oriented pharmaceutical industry are such laboratories as those of the Wellcome Foundation, which is financed by the parent industry, the Burroughs-Wellcome Corporation. Such companies plow back some of their profits into biomedical research with barely any strings attached. The resulting findings go not only to the parent companies but also freely into the open literature. These industrial foundations also have attracted prominent biomedical scientists.

The chief sources of new drugs have been, and are, the laboratories of the research-minded pharmaceutical industry. Of the hundreds of companies listed as pharmaceutical firms all over the world, not more than 40 possess facilities that produce the increasingly complex drugs that have transformed modern medicine. They are located primarily in Belgium, Canada, Denmark, France, Germany, Great Britain, Italy, Japan, Sweden, Switzerland, and the United States. Many of them are multinational corporations with production and sales organizations in many parts of the world. However, their research to discover and develop new drugs is usually done at their home bases unless tax, import, and monopoly laws favor a dispersion or even a duplication of effort in other countries.

WHY DRUGS ARE EXPENSIVE

In the pharmaceutical industry, the common goal of all scientists, engineers, and business people is to create and develop new drugs. At least some of these drugs must be profitable enough to repay the huge costs of their development and to make up for unsuccessful drugs. If we consider first the expensive path of study that followed the development of plant alkaloids into drugs, we see that it recurs in many other cases of medical research. A crude natural product or synthetic chemical is obtained, purified, and screened for biological activities. In the case of natural products, some idea of what effect to look for comes from folklore. As unreliable as such traditions are, they did, and sometimes still do, suggest better biological test methods than random screening.

When the chemical structure of a compound is known, as it is for most synthetic substances, further guidance as to what biological test might be fruitful comes from comparing it with structurally similar chemicals with known biological properties. Thus, if a new test compound contained a certain ring of carbon and other elements and if other compounds containing the same ring as a "skeleton" have lowered the blood pressure of laboratory animals, then the new compound might also lower blood pressure, and testing for this property would be in order.

The two hardest decisions for modern administrators of drug research and informed business executives alike are what specific research to start and when to stop it. They know that of every 5,000–10,000 compounds tested, only one makes it to market, often after 10–12 years of research and development. Suggestions and requests for novel drug therapies usually come from consulting physicians who discuss these possibilities with physicians within the company. The pharmacologists and other experimental biologists are consulted early to decide whether they can set up a suitable test to screen for the required activity. As the understanding of the biochemistry of some diseases improves, ideas for a new drug research project increasingly arise from participating scientists, biochemists, enzymologists, and molecular pharmacologists. If the proposed program requires the isolation and inexpensive manufacture of hormones or nutritional substances (e.g., essential amino acids or vitamins), organic chemists and fermentation experts in the company will have to judge the possibility of success for these operations.

A technique called recombinant DNA has permitted the commercial production of extremely complicated hormones and drugs, such as human insulin, by altering the genetic mechanisms of bacteria and yeasts so that they produce the desired compounds. Almost all major pharmaceutical research companies have started recombinant DNA experiments and have expanded their staffs by employing geneticists and other experts in these fields.

During the time of patent protection, most of the widely used effective drugs are highly priced. Developing a drug through preclinical and clinical trials requires that $220–270 million must be amortized during the period protected by the patent. Because the maximum length of patent protection is 22 years and the patent is usually issued during the preclinical work-up (which lasts about three years), one must subtract the time of clinical testing from the remaining 19 years. Clinical trials of average drugs take four to eight years. If we average this figure as six years, this leaves 16 years to recoup $250 million and to make a profit from which future investigations will be financed. At a modest rate of 10–12% of annual interest, the company has to chalk up a return of more than $10–12 million per year just to pay off the investment and overhead and to achieve a modest 3–5% dividend and a nest egg for future research.

Not many drugs provide such an income. Once in a while, however, certain new and revolutionary medicines score very high indeed. Hoffmann-LaRoche's diazepam (Valium), an anti-anxiety drug, became the best-seller among all prescription drugs worldwide and maintained this position for several years. Its sales were reported to add up to $1 billion per year but have declined now. Another best-seller was SmithKline Beecham Corporation's cimetidine (Tagamet), used principally to treat (and frequently cure) stomach and intestinal ulcers and pathologically high levels of stomach acid (*hyperchlorhydria*). During its peak years, its sales also may have amounted to $1 billion. A drug with similar activity, Glaxo Laboratories' ranitidine, also reached this mark. These exceptional cases bear witness to the sad fact that a large proportion of the world's population is beset with anxieties that are often a part of the vicious circle that leads to stomach ulcers, hence to more anxiety, and so forth.

Many other drugs for functional and infectious disorders have also attained respectable sales and profits. Derivatives of cephalosporin that are taken by mouth have often superseded injectable drugs against bacterial infections. Their sales amount to hundreds of millions of dollars. Drugs that lower the blood pressure, such as propranolol (a beta-adrenergic blocking agent) and the diuretics furosemide and hydrochlorothiazide, are in the same league. A mixture of levodopa and carbidopa, which has become the principal method to replenish deteriorating amounts of dopamine in brain cells of patients suffering from Parkinson's disease provides sales of more than $100 million per year.

Anything that sells for very large sums becomes an incentive for competitors, who try to capture a percentage of the market. After the expiration of a patent, some manufacturers and distributors of drugs produce and promote the original drug in its generic form. Several foreign chemical companies, especially in Italy and Turkey, now produce the raw materials of generic drugs. The purified high-quality powders or liquids are shipped to Western Europe or the United States, where they are made into tablets or coated pills.

Some pharmaceutical companies watching the profitable sales of a drug by other firms prefer not to wait until the original patent has expired. They want to market a competitive product that they hope will have some advantages over the original proprietary drug. Such substances may have a different chemical structure or biological action, but all too often they are only minor modifications of the drug they imitate. Even the best patent attorney cannot foresee all the little changes a competitor might make, and the most comprehensive testing of chemical variants still allows the chemicals of another firm to sidestep the claims of a drug patent.

Some people contemptuously refer to such competitive medicines as "me-too" drugs, but they forget that different products with virtually the same spectrum of biological activities have one real justification: No two patients are alike in their response to a given drug. A prescribed antihistamine may work on one person only to relieve allergies; it may make another person drowsy. These reactions may be reversed with yet another of the 30-odd antihistamines on the market. These differences are caused by the metabolic variations among patients and the workings of their immune defense systems. In any event, the availability of alternative similar drugs has important medical advantages, as physicians know. The choices allow doctors to find medicines to help their patients and produce a minimum of side effects.

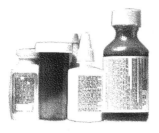

5

Modern Drug Discovery and Development

The first step in developing a new drug for a given health problem is to make plans for the chemical and biological work. A number of options are open to the participating scientists. The oldest option is screening, that is, subjecting a group of likely chemicals to a biological test in which some conditions of the clinical disease are imitated in laboratory animals. The requirement for success in screening programs is to develop a suitable, reliable, repeatable, biological test that is quick and not too expensive.

These requirements are most easily met in the design of drugs to treat infectious diseases. Bacteria, protozoans, rickettsias, and viruses are invaders foreign to the human or animal host. These organisms usually differ from higher animals in nutritional needs and internal chemical (*metabolic*) reactions. Thus, a chemical attack tailored to these microbes has a better-than-average chance of causing little harm to tissues and cells of the host. In addition, many disease-causing microbes (*pathogens*) can infect a number of hosts, including laboratory mice or rats. Thus, human and bovine tubercle bacilli infect mice; malarial parasites (*Plasmodium berghei* for mice and *P. knowlesi* for monkeys) are similar to those that cause human malarias; the human leprosy bacillus has been implanted into armadillos; and many other bacteria (such as staphylococci, gonococci, or streptococci) thrive in small laboratory animals. If a new chemical succeeds in killing such infections in mice, or at least in stopping their growth, without being too toxic to these animals, there is a good chance that the chemical may become an antibiotic for infected humans.

The greatest difficulties in developing drugs of this type are encountered with experimental viral infections because viruses need living cells for their life processes; they take over part of the host cells' metabolic functions and thus offer the chemotherapist or healer a mixed situation. Parts of the infected cells act like the virus; other parts retain the characteristics of the original cell; these actions are hard to separate. A chemical that subdues a virus often also subdues some of the life processes of the infected cells. As one wit said, "Flu is that disease that makes you feel sick six weeks after you get well."

The situation is similar in noninfectious diseases in which the metabolism of some body cells has deteriorated. Drugs that heal such disorders must restore the functions of the diseased cells without disturbing the normal body cells that surround them. At the least, such drugs should concentrate in the diseased organ but not elsewhere in the body. When one considers the circulation of the blood, this requirement is asking the near impossible, and yet it has been achieved time and time again. Animal tests that mimic a human disease often allow a drug to be selective in this manner, but the drug still does not work for humans. The reason for such failures is that no two animal species metabolize foodstuffs or foreign chemicals in exactly the same way. In some animals, a drug circulating through the blood or lymph vessels is held up by otherwise indifferent proteins, or it may pass through membranes that form a barrier in other animals. Some drugs are attacked and destroyed by certain body chemicals in one species but not in others. These differences mean that the metabolism of a drug has to be determined in at least two or three species before any predictions about its clinical behavior can be made. These obstacles must be overcome to determine whether the concentration of the drug in a body fluid—usually blood—is the same in a human as in an animal in which the drug works well. These measurements, carried out by refined and sensitive chemical analyses, can help pharmacologists choose among promising drugs for clinical trials.

CLINICAL TRIALS

Clinical trials form the most time-consuming, expensive, and unnerving part of the introduction of a new drug. Clinical trials are never done within the walls of the company that conducts the research. They are always farmed out to clinical pharmacologists and other specialists in internal medicine, often in university-affiliated hospitals. The company pays the bills for hospital beds, clinical apparatus, and tools; the salaries of physicians, nurses, orderlies, and test volunteers; and the expenses of patients who agree to try out the new drug. Typically, 25–100 patients are tested and the tests continue for two to four years. When the high overhead expenses charged by most hospitals are added, the bill for clinical trials quite commonly runs to $50 million or more.

After learning a drug's toxicity in several animal species, the clinical researchers must find the range of nontoxic doses in healthy human volunteers, with their informed consent. The drug may be administered by injection (*parenterally*) or by mouth (*orally*). The oral dose is usually larger because of the obstacles to absorption in the stomach and intestines. The effects of the drug on body weight and functions, individual organs, and the general health of the volunteers are recorded and reported to the U.S. Food and Drug Administration (FDA). Next come gingerly trials in patients suffering from diseases that the drug is intended to benefit.

In some instances patients with a given disease may be hard to find. For example, malaria has been almost wiped out in moderate climatic zones by killing with insecticides almost all the female *Anopheles* mosquitoes that carry this disease to people. It is therefore not practical to search for volunteer patients with malaria except in humid, tropical locations. In such cases, local volunteers must agree to be bitten by malaria-carrying mosquitoes so that the new drug may be tested on the course of their disease and, one hopes, may cure it.

To eliminate all bias in clinical trials, drugs are administered by double-blind methods. Only the principal investigator knows the dosage of a drug and which tablet or syringe actually contains the drug. Other tablets and syringes, looking alike, usually contain a *placebo*, that is, a preparation containing no medicine. Sometimes other look-alike forms may contain a different drug of known value in the disease under study. The doctors and nurses who give the drug are kept in the dark about the composition and strength of the doses so that their expectations cannot influence the patients' behavior. In this manner, statistically valid comparisons of the effects of the drug can be obtained, and such comparisons are demanded by the FDA before it approves a new drug application for general use.

Many other aspects of clinical trials cannot be included here, but a few inherent pitfalls follow. If the effectiveness of a drug cannot be established, all trials are stopped immediately because giving even a slightly toxic but ineffective substance can only endanger a patient without benefiting him or her. On the other hand, considerable toxicity is disregarded if the drug helps a life-threatening condition for which no other therapy is available. That is the case in some drug-resistant infections and also in cancers unless some alternative treatment (such as surgery or irradiation therapy) can be chosen.

While clinical trials are going on, animal experiments are continued to determine the long-term effects of the drug on every body organ. These long-term tests cannot be done in humans within the time allotted for the original clinical trials. In the meantime also, the manufacture must be scaled up from laboratory quantities to commercial production of kilograms or even tons of the drug. The length of time that the drug remains effective must be established and the manufacture of tablets, pills, and sterile solutions for injections worked out. These operations are not foolproof because mechanical compression, humidity, and heat may affect an otherwise stable compound. No organic compound is stable indefinitely, and therefore drugs should be discarded after their expiration dates.

The real evaluation of a drug by the medical profession gets under way as soon as the drug is approved and becomes available for general use. In this country, concerned doctors publish their findings in medical journals, and the FDA may or may not take these findings into account when changing its regulatory guidelines. In the United Kingdom and most European countries, physicians are required to send their observations to the agencies that control the use of drugs. Such observations are not all negative. To be sure, most physicians report unwanted or toxic side effects, but the significance of such phenomena must be evaluated statistically and not on the basis of scattered individual cases.

Clinical trials have become a jumble of scientific and legal requirements. The Delaney Amendment to the Pure Food and Drug Act rules that no food additive that causes cancer in any animal species may be tried in humans. The pharmaceutical industry has voluntarily extended this rule to drugs and routinely, before clinical trials, tests every drug in several animal species for any increase in the numbers of cancers found. This research is time-consuming and expensive, but it is necessary to avoid later litigation for malpractice. A similar situation holds for the *teratogenicity* of a compound, its potential ability to cause birth defects.

Teratogenicity is damage to genes. Its effects are pronounced in the early stages of fetal development and can result in death or deformed or poorly functioning offspring. A drug that is teratogenic in laboratory animals may or may not be so in humans. Teratogenicity for humans cannot be tested in controlled clinical trials because physicians cannot expose pregnant patients to that risk. Such damaging properties are only discovered later, after a drug has been in general use, and in such a case the drug is withdrawn immediately. Even though a drug has some beneficial activity (as did thalidomide, which was used to treat insomnia), men of all ages and postmenopausal women must be deprived of it. However, if such drugs were available in a medicine cabinet, they could not legally be withheld from young women who do not or will not read, or who will disregard a label that says, "Not to be used by women of child-bearing age."

An additional legal burden on clinical trials is the number and frequency of the reports on effectiveness, side effects, and toxic effects of the test drug that must be submitted to the FDA. This unproductive paperwork has alienated many clinical pharmacologists and has caused some multinational corporations to support clinical trials abroad, where the requirements are easier to fulfill.

During preclinical experiments and especially during careful clinical trials, observations of new and unexpected biological activities of a test drug may be recorded. For example, the drug imipramine, which was tested against psychoses, revealed its ability to relieve depression during clinical trials. In the same way, the diuretic hydrothiazides were found to lower blood pressure. Minoxidil, tested for effects on blood pressure, made hair grow on bald spots, and this cosmetic property has become its main use. Propranolol and other drugs (called beta blockers) for controlling rapid heartbeats (*cardiac fibrillation*) also lowered blood pressure, quieted the pain of angina pectoris, and even reduced fluid pressure in the eye from glaucoma. In such cases, medicinal chemists are encouraged to change the molecular structure of the compound in the hope of suppressing its original action and converting the newly discovered side effect to the main activity. Of course, this quest requires going back to home base and taking the new substance through all preclinical and clinical stages. However, this method is valuable for designing new drugs.

If one of the side effects of a drug is prominent enough to deserve immediate clinical application, the FDA amends its approval of the use of the drug to include the newly discovered use. This amendment often takes considerable time because evidence and paperwork must wind their way through bureaucratic channels. These delays, which deprive patients of valuable help, have been criticized. However, effectiveness and lack of toxicity at the dose levels necessary for the new activity are the overriding considerations for extending drug approval. Specialists in geriatric diseases have pointed out that one of the principal characteristics of aging is a slowing of the metabolism. Food and drugs alike are metabolized more slowly in older animals and humans. Therefore, geriatric patients should be prescribed different doses of drugs as compared to patients of middle age, and certain drugs that depend on metabolic chemical activation should be avoided in older patients.

OVER-THE-COUNTER DRUGS

People suffering from minor ailments usually do not want to see physicians about their complaints. They go to drugstores and scan the shelves for medications that promise to deal with their trouble. If they are unsure what to choose, they may seek (free) advice from pharmacists and act accordingly.

Few over-the-counter (OTC) medications contain only one active ingredient; most of them are mixtures of several compounds designed to serve several purposes. Cold medicines may serve as an example. Because no drug is known yet to inhibit the virus that causes a cold, two or more drugs are put together to counteract and reduce the symptoms of a cold, not its cause. An antihistamine is added to alleviate sneezing, runny nose, and other allergic manifestations. Antihistamines produce a sleepy feeling as a depressant side effect, so caffeine is added to make the patient alert. A vasoconstrictor is added to contract blood vessels and thereby make room for the expansion of nasal airways so that the patient can breathe freely.

Another example may be found in OTC pain-relievers, or *analgesics*. Several analgesics contain aspirin, some contain acetaminophen, and many contain ibuprofen as the active ingredient. If the drug is meant to subdue a headache involving the sinuses, the analgesic is compounded with a vasoconstrictor. If the drug is meant to help the patient sleep soundly through pain, a depressant antihistaminic is added, most often diphenhydramine, which is the active ingredient of many sleeping agents as well.

In this volume, the direct intervention of a discomfort or disease by a given drug is described. Mixed medications as one finds them so often in OTC drugs are not covered except in cases in which each ingredient of a mixture enhances the total effect on the disease by different and cooperative mechanisms.

HOW DRUGS ARE MADE

The earliest (but still widely practiced) method of preparing drugs is the extraction of natural products, such as plants, animal tissues, marine organisms, or microbial forms of life. Extraction with aqueous or organic solvents is followed by concentration, separation of mixtures, and purification, often with the aid of modern laboratory instruments and machines. These processes are first performed in small laboratory batches. If the purified material is biologically active and has promising therapeutic properties, the procedures can be stepped up to factory-size quantities.

Most medicinal drugs are made by synthesis from simpler organic reagents. Developing this process calls on the intellectual acuity, expertise, and vision of the chemist. Starting with pots and kettles filled with relatively simple chemicals, reagent after reagent is added. Each resulting intermediate is purified, again made to react with some pertinent compound by heating or cooling, stirring, distilling, fractionating, and purifying until the desired product is obtained. In the special case of chiral compounds (*see* the last section in this chapter), additional separations are required. This type of synthesis has sometimes been aided and simplified by processes labeled *biotechnology*.

BIOTECHNOLOGY IN DRUG PREPARATION

Biotechnology applied to chemistry involves essentially letting living biological organisms or their enzymes do the work usually performed by chemists. Perhaps the most ancient example of chemical biotechnology is fermentation of alcohol. Starch or other carbohydrates are fermented by yeast to furnish dilute aqueous solutions of ethanol (*mash*), from which ethyl alcohol is recovered by fractional distillation. Alcoholic products with different flavors can be obtained in this way, depending on what is added to the fermentation mixtures. Scotch whiskeys, gin, cognac, moonshine, and pure ethanol are distilled out of such fermentation vats. Wine is made similarly, using the sugars of grapes as the materials to be fermented.

Alcoholic fermentation has also taught us that living microbes such as yeast are not needed to convert carbohydrates to alcohol. An enzyme, that is, a catalytic protein isolated from yeast, performs the fermentation. This process established once and for all the fact that cellular chemicals can catalyze chemical reactions without the need for living organisms.

Hundreds of bacteria, molds, fungi, and other microbes can ferment different starting materials (*substrates*) to furnish specific chemicals, e.g., antibiotics. One can seed such fermentation mixtures with all kinds of pure organic compounds and thus alter the resulting fermentation products. The same microbe produces different analogous or related biochemicals depending on what one adds to the brew. This process amounts to molecular modification by biotechnological means.

If a microbe can metabolize a given chemical, scientists can also use this property to biosynthesize some complex substances, often decreasing the number of chemical operations they would have to perform in the laboratory to achieve the same results.

In some instances, it has been possible to alter the genetic make-up of a microbe and force it to biosynthesize substances it could not produce without this genetic change. An example is the production of human insulin by common colon bacteria that have been altered to do the work ordinarily carried out by the beta cells of the islets of Langerhans in human pancreatic tissues. It is no longer necessary to slaughter animals to collect their pancreases; today we use instead microbial fermentation to obtain the human hormone, not the animal hormone with its inherent allergenicity.

GETTING DRUGS READY FOR USE

Most of the commonly used drugs are solids that can be pressed into tablets; liquid products are usually enclosed in plastic capsules. The plastic is chosen from polymers that dissolve in the acid stomach or in the slightly alkaline small intestine. Tablets and capsules are polished, coated with flavoring substances, and colored by machines that weigh the exact amount of drug, shape the powders, enfold the liquids, coat the pills, and get them ready for bottling and distribution.

Inevitably, some of the tablets or capsules are imperfect, perhaps with a broken corner or some other fault. Such damaged units must be removed before a given batch is released for sale. This removal is done by visual inspection. In a typical case, three workers sat hour after hour, day after day, watching thousands of capsules float by on a slowly moving delivery belt. Any imperfect capsule was rejected manually by stopping the belt and removing the damaged pill. The failure rate for detecting damaged capsules was low, estimated at one in a million. But this was an expensive operation. The workers sat without spotting a faulty capsule for hours at a time, while their wages added up.

In discussing this problem, an experimental psychologist made a novel suggestion. He had been employed to improve animal tests for potential drugs for mental disorders and had worked out methods that were based on conditioned responses. In these tests, an animal is exposed to a stressful stimulus and tries to escape the pain or stress. One can teach an animal to overcome its fear, at least partially, by rewarding it with food and water each time it withstands more stress.

The experimental psychologist suggested a similar procedure for inspecting the capsules. Pigeons were taught to watch the stream of capsules floating by and to peck a lever every time they saw a faulty capsule. Each correct peck on the lever was rewarded by a grain of corn or a drop of water. The running belts were adjusted to remove any faulty capsules. After a training period, the pigeons took over the duties of the inspecting people. This method worked so well that oversight rates fell from those previously recorded by the human inspectors.

This should have been the end of the story, but a new difficulty arose. The displaced workers went to their labor union representatives and asked that their previous jobs be reinstated. The company had no choice but to retire the trained pigeons and to reinstall the human inspectors at the running belt.

CHIRAL DRUGS

A person's two hands, placed palm down on top of each other, cannot be superimposed. The left hand's thumb protrudes to the right, the

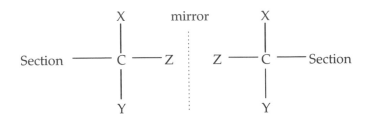

Figure 1. Two compounds that are enantiomers are at some point mirror images of each other.

right hand's thumb points to the left. If one makes the palms of the hands face each other, the right hand is the mirror image of the left hand. Human hands are not symmetrical. The Greek word for hand is *chiros*. Many molecules are also nonsymmetrical; they are *chiral*. Two molecules that differ at some point that makes them asymmetric are called *enantiomers* (Figure 1). A 50–50 mixture of two such enantiomeric forms is called a *racemic* mixture, or for short, a *racemate*.

Because enzymes are proteins composed of chiral amino acids, reactions and syntheses catalyzed by enzymes lead to chiral compounds if the chemical structure of the reaction products involves the occurrence of enantiomers. In many cases, this catalysis is critical for the success of getting the desired isomer without tedious and protracted chemical separations from a product obtained by conventional chemical synthesis.

Drug molecules must fit an enzyme or receptor to bind there snugly and initiate a series of rapid chemical reactions that result in a biological effect. The enzymes and receptors are asymmetric and therefore bind to a drug racemate at often different rates. If these differences are small, the two enantiomers may have almost equal biological activity. If the differences are large, only one enantiomer may be active, or one may be toxic or even have a different kind of biological activity.

If one of the enantiomers is toxic, it obviously must not be given to a patient. If the enantiomer has a different kind of activity, it should be removed because a patient should not be given any unnecessary medication. The FDA has recommended that in the future all asymmetric drugs should be prepared and administered as single substances, not mixed with enantiomers that have no effect on the disease or have no activity. Methods to produce only one enantiomer are

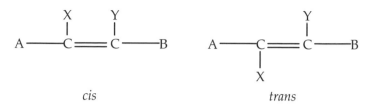

Figure 2. Geometric isomers.

known, and this problem can therefore be resolved, although usually at a high cost.

A second type of structurally closely related compounds is called *geometric isomers*. They contain a rigid bond or section such as the double bond shown in Figure 2. The two portions cannot rotate around this rigid bond. The compound with X and Y on the same side of the immovable bond is called *cis* ("here"), the other one *trans* ("there"). Cis and trans pairs behave differently when they approach an unsymmetrical enzyme or other receptor. One of them may bind snugly, whereas the other one may not fit. Therefore, one of the two geometric isomers is usually more biologically active than the other. Researchers are trying to use these properties to their advantage in designing new drugs.

6

Molecular Modification of Prototype Drugs

Discovery of a totally new chemical with a novel biological activity is such a long-range gamble that the drug industry is forced into short term efforts to furnish drugs with moderate to good potential value. The search for look-alike drugs is an example of short-term efforts that are a reasonable undertaking. Indeed, the majority of chemists in the pharmaceutical industry synthesize and study such drugs by a process called molecular modification. Because of the importance of this work, a few explanations of their techniques are in order.

A chemist does not set out to make a drug that acts exactly like one already known, with all its advantages and disadvantages. Instead, investigators try to overcome the disadvantages, usually unwanted side effects and signs of toxicity. Perhaps the new substance should be more powerful so that a smaller dose can be given, presumably with fewer side effects because less drug will be present in the body.

If two different chemicals cause a similar effect, they probably react at the same chemical positions or receptors on the target cells. For this purpose they must fit snugly to these receptors. Picture a glove that fits your left hand. Another person of similar size may try on one of your gloves and find that the left hand will fit your glove, but neither your right hand nor that of your friend will fit it. In other words, a similar shape of the two biologically active molecules is required for similar biological action.

Similarity of shape, though, is not enough. There must be some inducement for the molecules to approach the receptors and bind to them. Chemical attraction between two molecules—drug and receptor—is caused by electrons. Electrons are negatively charged and therefore are attracted to positive poles or places; they are *nucleophilic*. When an electron is removed, the portion of an electrically neutral molecule that stays behind has lost a negative charge and so becomes positively charged. Such a positive ion then tries to become neutral and at rest by joining an electron-rich, negatively charged particle somewhere else; it is *electrophilic*. Although this account is an oversimplification of chemical reactions, it may serve to illustrate some essentials of chemical similarity.

For two different chemicals to react at the same receptor, they must have similar shapes and similar electronic behavior. This behavior is also called similar *binding* ability because binding comes about by joining or transfer of electrons. Two molecules that have similar shape and similar electronic behavior are called *isosteres* (from the Greek: *isos*, meaning like and *stereos*, meaning shape); this designation also applies to similar portions of molecules. Two chemically and biologically similar compounds are called *bioanalogues*.

Medicinal chemists have worked out a set of rules and limitations for this property, and they apply them to drugs that they want to imitate. Following these guidelines usually results in a number of similar compounds (*analogues*), but only a few of them are suitable for further study. Studying all of the analogues is a great waste of time and effort because all the unsuitable analogues are discarded.

Computer-minded chemists have tried to limit the numbers of almost-right compounds by measuring various properties of a set of chemicals and having the computer pit one property against another. This method helps occasionally and is cheaper than wasting months of laboratory time. Therefore, almost all research companies have installed computers for that kind of work. On the whole, changing molecules bit by bit remains an art in which experience, luck, and clever experimentation go hand in hand with sophisticated calculations of electron clouds and shapes of molecules as determined by X-ray diffraction. It is hard to decide how to improve upon the structure of a useful drug. To do so, one has to work within the limits of shape and electrical affinity and try one's luck time and again.

The life processes of animals, including human beings, are caused and regulated by tens of thousands, perhaps more, of chemical

reactions. Many kinds of body chemicals have been isolated and identified, and some of their usually multiple functions have been described. But innumerable puzzles remain. It is as if the yarn needed to
make a multicolored rug had been spread on the floor, and the
weaver explained that people would walk on the rug after it was
made but failed to say how the weaving and blending of the different
strands would be done. Perhaps he or she would decide details while
weaving. In life, the actual interactions of the spectrum of biochemical
"yarns" found in an animal organism are difficult to determine
because methods for studying them in their natural environment are
very complicated.

The Harvard psychiatrist Seymour Kety has illustrated this
dilemma. He says that a delegation of extraterrestrial beings came to
visit Earth and wanted to learn what makes humans tick. They
decided on a biochemical analysis of the Earth creatures. A few hun-
dred people were caught and ground up in a giant meat grinder, and
the material was extracted, fractionated, and identified chemically.
The aliens found dozens of amino acids, hormones, nucleosides, and
many other compounds. They arranged this mixture in what they
thought was a logical sequence and nodded wisely. "Now we know,"
one of them said, "how the humans composed Beethoven's Ninth
Symphony; wrote Shakespeare's *Hamlet*, Goethe's *Faust*, the Koran,
and the Bible; and built the Golden Gate Bridge and the Taj Mahal."
We are at the same stage of interpreting life processes.

With the coming of very sensitive analytical instruments in the
past 40 years, scientists have been able to study ever smaller amounts
of hormones, parts of the immune system, tumor-causing viruses,
fragments of nucleic acids and genes, traces of metal ions, and other
elusive biochemicals. At first, progress was slow because the sheer
learning of the new methods took years and all the difficulties that
bedevil complicated new enterprises had to be ironed out. Lately the
pace has sped up, and the new methods are spreading to such fields
as genetics and psychology, with startling results. Scientists are even
beginning to explain such biological mysteries as memory in terms of
chemical reactions.

When a drug is administered, it intrudes into this interwoven,
ever-changing, and unbelievably complex array of biochemical reactions and is supposed to restore "normality," or *homeostasis*, the name
given by the physiologist Cannon to a healthy interplay of all the fac-

tors that keep the body running smoothly. Yet using a drug is a bit like sweeping a broom through a disorganized mass of litter while trying to line up the pieces in an orderly way. The nature of homeostatic mixtures of chemicals or of their disorganized states in disease is difficult to unravel. Now at least we know about many individual chemicals in the mixtures, although many of their reactions still baffle us.

Fortunately, the individual puzzle pieces can be mentally and chemically rearranged, much as molecular modifications form the basis of designing similar drugs. Molecular modification has already been applied to hundreds of products called *metabolites*, which have been fished out of the mixture of biochemicals in living organisms. The reasoning behind these experiments is that nature has not necessarily evolved the best chemicals for a given purpose, but often only the most chemically convenient. For example, cortisone and hydrocortisone are found among similar compounds in extracts of the outer layer or cortex of the adrenal gland. These two hormones help to counteract inflammation and arthritis, but they are not the best substances for this purpose. By changing these molecules, medicinal chemists have synthesized similar analogues (such as prednisolone and dexamethasone) that are highly useful drugs with stronger and more selective action on inflammation.

Two examples of successful molecular modifications of metabolites are para-aminosalicylic acid and methyldopa. Salicylic acid was believed to support oxidation of tubercle bacilli. *Para*-aminobenzoic acid (PABA) is a general growth factor (*vitamin*) for bacteria that make folic acid, a product biosynthesized from PABA. By combining salicylic acid and PABA in one molecule, chemists produced a drug called para-aminosalicylic acid, which has become an effective treatment for human tuberculosis.

The other example is based on an amino acid called dopa for short, which serves as a precursor for dopamine. Dopamine in turn is converted further into epinephrine (or adrenaline), which causes some types of high blood pressure. A small CH_2 group was added to dopa to make methyldopa, which blocks the chemical pathway to epinephrine. This block is one way to control blood pressure. High blood pressure can now be reduced to more normal levels.

The most successful use of modified metabolites as drugs has been to combat foreign or malignant invasive cells. Cancer cells divide in an uncontrolled, often rapid fashion. Because each new daughter cell needs a cell nucleus, nuclear material—especially DNA and RNA types of nucleic acids—is constantly manufactured in these cells. These nucleic acids contain several basic units called purines and pyrimidines. Two of these are adenine, a purine derivative, and uracil, a pyrimidine derivative. By making a slight change in adenine, one obtains 6-mercaptopurine, which enters the chemical conveyor belt leading to a nucleic acid but, being a faulty building block, prevents the construction of a working nucleic acid and thereby of new cell nuclei.

Another of these pyrimidines is called thymine. It has a small methyl group (CH_3) attached to the pyrimidine portion. When this methyl group is replaced synthetically by a fluorine atom, F, one gets 5-fluoruracil, which likewise prevents DNA from functioning in forming a new cancer cell nucleus. Physicians use these two drugs as powerful *antimetabolites* for chemotherapy to treat many cancers in addition to or instead of surgery or radiation.

The chemical methods for molecular modification are also used in planning drugs to counteract metabolites. The best part of trying to develop a new metabolite *antagonist*, as these drugs are called, is that the chemist need not imitate a competitor's drug but instead uses as a starting point natural substances made by the body—a much more satisfying scientific motivation.

Whether researchers try to develop new drugs by modifying prototype drugs or chemicals that are found in nature, their goals are the same: to produce drugs that do the job for which they are designed with few side effects.

7

Drug Use and Abuse

Some people pride themselves on never taking any drugs whatsoever. This abstinence is to be recommended during pregnancy because some drugs can traverse the placenta and may cause adverse effects in the fetus. In most conditions, however, not taking a drug when it might alleviate a disease or discomfort is unjustifiable. Religious fanatics may deny themselves the comfort of medications, but rational individuals should never fall into that trap.

The other extreme encompasses people who take too many drugs. Some people take every drug in sight, and others take a drug because it was given to them free or sold as a bargain. This statement is true not only for vitamins, although they are first on the list. One elderly woman lined up 24 capsules and tablets on the breakfast table and washed these pretty, multicolored chemicals down with another drug product, coffee. She had collected her medications quite simply. Her physician prescribed three or four for her, but when she bought them she was assailed by doubts. Would they really work? Would it not be better to ask for a second opinion about her condition? Insurance would pay most of her bill, and so she saw a second—and finally a third—physician. These doctors diagnosed the same illness as the first one but prescribed different brands of the same medication. They also recommended a few vitamin capsules, and that is how she ended up with all those pills. The outcome was that she ingested three times the therapeutic dose of each medication. It is no wonder that she landed in a hospital with toxicity symptoms from these overdoses.

This case is not unusual. The high cost of visits to a physician drives thousands of patients everywhere to attempt self-medication with over-the-counter, nonprescription drugs. Catastrophic intoxication is not more prevalent only because over-the-counter drugs are often not very effective. This ineffectiveness, though, does not protect sensitive individuals from allergic reactions to drugs. Nor does it eliminate the danger of overmedication in attempts to commit suicide.

Many patients do not comply with their physicians' prescriptions and directions. Some patients will not take prescribed drugs at all, for whatever reason—unwillingness to obey orders; adherence to Christian Science, which shuns drugs; or because their bridge partner or beauty parlor operator has warned them of side effects of the drug. Other people want to save the expense of buying prescriptions—and some drugs are very expensive.

DRUG INTERACTIONS

Physicians are aware of drug interactions, but in years past they did not adequately focus on such dangers. The best advice on drug interactions can be obtained from a knowledgeable pharmacist, some of whom now hold Doctor of Pharmacy degrees. A pharmacist who can supply helpful information can be as important to the patient as a physician.

Drugs may interact with each other by influencing each other's metabolism or absorption. For example, phenobarbital can stimulate the oxidative destruction of several unrelated drugs and thereby make them less long-lasting and less effective. Anticoagulants are made less effective in combination with barbiturates and a spate of other, unrelated drugs. When such drugs are used concurrently, blood clotting may occur. It is therefore imperative to follow a physician's careful directions in such cases. It is also important that patients tell physicians about any medications they are currently taking, especially something prescribed by another physician.

Two drugs that have the same type of activity should not be taken simultaneously unless toxicity from large doses of either of these drugs is to be minimized. Many patients heap one painkiller on top of another; this action may induce depression and other unwanted side effects. Two drugs that have opposite activity should not be administered together either. Thus, vitamin K, which promotes blood coagulation, nullifies the action of anticoagulants.

Not all harmful interactions arise from drugs that interfere with each other; dietary factors also play a role. The most famous case is that of cheddar cheese, red wine, or bananas and the antidepressant monoamine oxidase inhibitors. These foods contain a biogenic amine called tyramine, which raises the blood pressure. Ordinarily, tyramine is destroyed rapidly by the enzyme monoamine oxidase, but if this enzyme is held in check by the monoamine oxidase inhibitors, tyramine may lead to a dangerous rise in blood pressure. Likewise, while taking an antibiotic, it would be foolish to include vitamins in one's diet because they promote the essential life processes of the pathogens. Tetracycline loses its antibacterial activity when it is taken with milk or other calcium-containing foods because it forms an insoluble product with calcium.

On the other hand, combining several drugs in one tablet may be not only convenient but also pharmacologically justified. Many anticancer drugs must be taken in doses verging on overt toxicity if they are to be effective. By mixing lower doses of three anticancer drugs and attacking tumor growth by three different biological routes, some of the toxic side effects of the individual components of the mixture can be minimized.

Some drugs, when taken in excessive doses or over long periods of time, induce symptoms of chronic toxicity that may be called drug-induced diseases. A considerable percentage of patients seeking admission to hospitals are victims of such drug-induced diseases.

Reversing the chronic effects of a drug is difficult if not impossible. This difficulty can be seen even in habituation to such an innocuous drug as caffeine. The British people consume 4.5 kg of tea per person annually, and Americans drink 8 kg of coffee per person per year. These figures represent enough caffeine to affect profoundly heart rate, diuresis, and respiration. Yet coffee and tea are drunk mainly to provide stimulation of the central nervous system.

Because all drugs are tested for chronic toxicity in laboratory animals before being admitted to clinical usage, experienced physicians are able to adjust prescriptions in a manner that promises to avoid chronic toxicity.

PHYSICIANS AND DRUGS

The physician is regarded by patients as a dual person. He or she must be competent in diagnosing and treating diseases and in prescribing the most up-to-date and effective and the least toxic drugs. Most patients require a second quality in a doctor, whether as a primary family practitioner or a specialist. They want him or her to be a sympathetic, comforting parent figure, emanating professional authority and compassion. We demand a good deal from a physician who sees and treats 20 and sometimes 50 or more patients every day. However, it is the physician's duty to explain the diagnosis to the patient, prescribe adequate medication, and advise patients about side effects or even failures of the drug prescribed. The physician should also make it clear when and how long a drug should be taken. Patients with little education may stop taking an antibiotic as soon as they are free from fever even though they invite a relapse if they interrupt the treatment before all pathogens have been killed. A factual attitude toward all drugs without prejudice does much to help in the drug therapy of a disease. This factual attitude should include the hope that a given drug will do the task expected of it, but that the drugs, like all the creations of human brains, cannot be expected to be perfect.

A physician's decision in such matters must include risk assessment. As an aid to risk assessment, the manufacturer of a drug furnishes a drug insert, a comprehensive description of desired effects and of undesirable or dangerous side effects. Carcinogenic danger and teratogenic properties (causation of birth defects) must have been determined in animals before the drug could be admitted by the FDA for medical use. On the other hand, a patient should not become discouraged when he or she reads these many drawbacks. The manufacturer lists all pertinent side effects, even if they have been observed in only one patient out of 10,000 or more. The pharmaceutical industry uses this policy to protect itself from malpractice claims. If a malpractice claim is made, the defendant manufacturer can point to the drug insert that accompanied the medication and say, "We warned you."

The enormous monetary awards allotted by American juries in malpractice cases induce patients and their lawyers to gamble on winning such suits. Not only the drug manufacturer but also the prescribing physician has to ward off legal actions by taking out extensive and very expensive malpractice insurance policies. In turn, the cost of these policies is charged to the patients or their health insurance companies, and these costs raise their premiums to account for the actuary expenses of their business.

If we add the cost of all this legal action to the high costs of complicated chemicals, of the grueling intellectual research by scientists and clinical physicians in research hospitals, of manufacturing the drug by skilled labor, and of the expenses of distribution, and we factor in the often short market life of a drug, we can understand why drugs are expensive.

QUACKERY

The horse-drawn medicine cart, gaily painted and decorated, was a cherished American tradition in the same league as the circus coming to town. The driver of the cart was a combination comedian, huckster, self-appointed health adviser, and fake. Crowds greeted his visit, and men and women alike succumbed to the hard-sell advertising of snake-bite oil, hair tonics for balding heads, rejuvenating nostrums, aphrodisiacs, deworming medicines, and backache liniments.

The blandishments of these unauthorized "doctors" and the ineffectiveness of their medications were the direct cause of the Pure Food and Drug Act that the U.S. Congress passed in 1906. This legislation abolished most, although not all, outright quackery that had been foisted on the public by medical swindlers in a shameless manner for centuries.

Control of any new legislation takes time, and even today, almost 90 years after it was enacted, the public is still confronted by a few unscrupulous manipulators who attempt to extort money from sick and despairing patients and their families. The advertisements of pain relievers in the media are now couched in careful sentences designed to stay on the razor edge of legally allowed language, but the average listener, suffering from pain, often does not see through the line dividing truth from deceit. The same holds true for the ointments that are said to relieve the pain and itching caused by hemorrhoids, the adhesive that holds dentures firmly in place, and all the other concoctions that pay for the broadcasting time of the evening news.

Such extravagances are of minor importance, and reasonable people smile at them and chalk them up to the perverse desire to believe in something that cannot be true. The Romans had a proverb for that: *Mundus vult decipi, ergo decipiatur* ("The world wants to be deceived, therefore let it be deceived"). Only when health or recovery from a disease is threatened by the acceptance of fraudulent therapeutic methods does the medical profession have to draw a line.

The dread of cancer has driven desperate patients and their families into the hands of would-be healers and the use of unproven medications. Now that early diagnosis and combination treatment by irradiation therapy, surgery, and chemotherapy have greatly reduced morbidity and mortality from malignancies, this fear should have been replaced by cautious optimism. But the two following incidents of serious quackery do not support such enlightened attitudes. The first of these incidents was the announcement by two Balkan "healers" about 1952 that they had discovered an anticancer agent in horse serum. They manufactured the product, named it Krebiozen (Krebs means cancer in German), and allied themselves with a prominent physiologist who, in an as-yet unexplained delusion, gave the drug respectability. There is no need to state the fact that Krebiozen was a hoax and did no good.

The second incident was the case of laetrile. This material was obtained from crushed apricot pits. They were extracted, and the evaporated extract left behind the agent. Chemical studies revealed that it contained a glycoside of mandelonitrile. This known compound can split off highly poisonous hydrogen cyanide, but in very small amounts. Mandelonitrile and its glucoside have no known therapeutic action; hydrogen cyanide is a general cell poison that may have accounted for some of the toxic symptoms caused by laetrile. Although this information was widely disseminated and supported by a careful study by scientists of the National Institutes of Health (NIH), thousands of deluded cancer patients insisted on taking the drug. Laetrile was banned in many states, but patients traveled to Mexico, where it remained available. Politicians in several localities were threatened with reprisals if they refused to reintroduce laetrile. Patients who believed in the power of this agent, though, refused to submit to approved medical methods of cancer treatment and thereby hastened their own demise.

Acquired immune deficiency syndrome (AIDS) is an as-yet incurable slow-onset disease that is transmitted by contact with body fluids of infected individuals. Infection with the causative virus (the human immunodeficiency virus, or HIV) may be brought about by sexual contact, by blood transfusion, or by contamination with blood products. Despair, especially among sexually active persons, has led to experimentation with indefensible quackery methods of treatment. "Energy soups," mushroom extracts, bee pollen, massage and meditation therapies, herbs to boost the damaged immune system, and even thymus cells in bottles are being offered to desperate victims by fringe practitioners and ruthless profiteers.

A curious invitation to quackery has sprung up in scattered groups of mostly young people who are seeking security in an insecure world. They try to lead a pioneerlike primitive life. Among their habits is the exclusive consumption of "natural" foods, which they grow themselves or buy in health food stores. As long as these stores offer a diet of nuts and vegetables, one might regard their wares as refreshing variations from prepared, prepackaged, and precooked meals offered by supermarkets. But the shelves of health food stores also hold materials promoted as natural drugs and sources of essential nutrients, minerals, and vitamins. They are botanical powders, crystals, oils, and extracts, and some of them are indistinguishable from similar products offered in oriental markets. A number of these products, however, are totally worthless. They must have slipped by the inspection of government agencies that watch over truth in advertising. This situation would not matter if the customers of these stores had the facts about such products at their fingertips, but many people are uneducated and gullible. Such wasted purchases can only be prevented by more effective consumer protection.

The barrier to dangerous and counterproductive quackery is education of the lay public. Constant reports of suspected cases of quackery, backed by scientific evidence translated into plain English, can dispel superstition and ignorance bordering on voodoo. With a wider emphasis on science in the public schools, the next generation should take a fresh look at beliefs that have come down through the ages, before understanding and searching for the truth had reached the state that we now value. In medicine there is always an outside chance that some natural or synthetic product may turn up as a new and valid therapeutic agent. Until such a claim has been verified by preclinical and approved clinical studies, however, the public should beware and remain skeptical.

The best way to judge the potential therapeutic value of a nonprescription drug preparation is to read the label specifying its contents. There is usually a generic name for the active ingredient that the lay person may not know. Consumers should ask pharmacists whether the generic products will benefit their conditions. If the pharmacists say that they will, the consumers may take the medications with confidence but should carefully observe dose levels, timing, and other directions on the labels. The rest of the list of contents may be disregarded.

The world will never be rid of the quacks and profiteers who take advantage of people who are in pain or need. It is up to consumers to protect themselves from these opportunists.

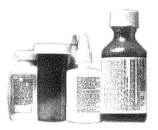

8

Neurohormones and Drugs That Affect the Central Nervous System

NEUROHORMONES

Ever since people began to use alcohol, opium, cannabis (in marijuana and hashish), tobacco, and peyote, drugs that alter mental states have been regarded with such emotions as awe, fear, and disgust. Some of these substances cause detachment and withdrawal in the people who ingest or smoke them or result in sleep or indifference to outside influences. For centuries, surgeons used alcohol and opium to blunt the intolerable pain of amputations and other operations. Some drugs were used in religious or orgiastic rites in which the participants wanted to communicate with deities or to shed restrictions taught by their societies.

In lower doses than those causing drunkenness or completely drugged behavior, alcohol and marijuana can improve sociability and provide a feeling of success. A sniff of ether may produce overexcitement, and barbiturates and other *hypnotics* (sleep inducers) act as sedatives, making animals and humans go to sleep. Such sedative-hypnotics are thought to have been the earliest drugs used for depressing brain activity. At the other end of the spectrum, cocaine and ephedrine were among the early brain stimulants that could overcome sleepiness, increase wakeful awareness, and improve the ability to perform complex tasks. Later, amphetamine, methamphetamine, and some other derivatives of these drugs became widely used but were soon banned as dangerous, addictive agents.

In 1867, one of the founders of organic chemistry, Adolf von Baeyer (1835–1917), first synthesized a laboratory curiosity called acetylcholine. He had no idea that he had described an important neurohormone. In 1900, the American physiologist Reid Hunt (1870–1948) studied the effects of acetylcholine and found that it depressed blood pressure in laboratory animals. Because this property promised to reduce high blood pressure in humans, extensive pharmacological work followed. Two contrasting types of activity of acetylcholine were found. One activity resembled that of nicotine and was called nicotinic. The other activity was similar to that of the mushroom poison muscarine; it was called muscarinic. Muscarine had been identified as the highly toxic substance in fly agaric, *Amanita muscaria*, the familiar red mushrooms with white spots that are often illustrated in children's books. Because of these different properties of acetylcholine, these activities were named cholinergic. Antispasmodic drugs such as atropine block the muscarinic component of cholinergic activities but have little impact on the nicotinic action of acetylcholine.

Each involuntary (*autonomic*) muscle and gland is "wired" (*innervated*) by two types of nerves: *sympathetic* (or *adrenergic*) and *parasympathetic* (or *cholinergic*) nerves. Any two nerve cells are joined by a fine network of fibers (*dendrites*), which are enmeshed in a junction called a *synapse*. Because the dendrites of one nerve cell do not touch those of the next one, a tiny space separates them. Bridging the gap is done by chemicals that enable a feeble electric current of microvoltage to carry an impulse across the synapse. Several types of chemicals are involved. One chemical is called a neurotransmitter, another one, a neurotransmitter transporter. The transporter chemicals are complex proteins with a sugarlike component (*glycoproteins*). Transporter chemicals help to pump the neurotransmitter back into storage cells when it is no longer needed. The other chemicals, the neurotransmitters, are simpler organic compounds than glycoproteins. Four of them have been studied widely: dopamine, epinephrine, and norepinephrine on the adrenergic side and acetylcholine on the cholinergic side.

The cholinergic transmitter is acetylcholine. In the sympathetic nervous system, the principal transmitter is norepinephrine (or *noradrenaline*); between the tip of a sympathetic nerve and the "end organ" that is stimulated (the muscle or gland) the transmitter is acetylcholine. In the parasympathetic system, the transmitter is acetylcholine at all points. Other neurohormones have also been recognized, particularly serotonin and simple amino acids such as gamma-aminobutyric acid (GABA). Serotonin is 5-hydroxytryptamine (or 5-HT) and acts at special receptors. These compounds, in conjunction with the transporters, initiate a chain of events leading to the transmission of impulses across the synapse. Some of them appear to play specialized health roles, such as lifting depression by serotonin and blunting pain by the body's own enkephalin peptides.

An electrical signal has to be amplified as it travels along a nerve; this amplification is accomplished by creating inorganic sodium, potassium, and calcium ions—that is, electrically charged atoms—from nonionized salts. In turn, the activation of these ions is performed by stepwise ionization of the transmitters and modulators (transporters) just described. This is a greatly simplified picture of the way nervous impulses flow, but it should suffice to guide us in discussions of drugs that stimulate, depress, or disorganize the nervous system. Some of these drugs have freed us from age-old prejudices and have improved the treatment of mental and nervous diseases beyond the wildest expectations of only five decades ago.

The first biochemicals recognized as neurotransmitters, acetylcholine and norepinephrine, faced a number of hurdles initially. Norepinephrine especially was an afterthought to the discovery of epinephrine (adrenaline) at the turn of the century. Epinephrine was found in human adrenal glands, which are located at the top and back of the kidneys. The glands consist of an inner layer (the adrenal medulla) and an outer layer (the adrenal cortex). The adrenal medulla secretes epinephrine; the cortex secretes cortisone, hydrocortisone (cortisol), and many similar steroid hormones. Among a number of other functions, epinephrine constricts blood vessels, increases blood pressure, stimulates the heart, and raises the concentration of sugar in the blood. It was relatively simple to synthesize epinephrine in the laboratory. In the course of this work, its slightly smaller first cousin, norepinephrine, was also prepared, although its role in nervous transmission was not explained for another 40 years.

Before 1950, ephedrine and amphetamine to stimulate and barbiturates to sedate were the only important substances available to
doctors for treating mental conditions. In 1952, three different drugs
made their appearance and opened up the era of true therapeutic
drugs in psychiatry. They were chlorpromazine, the original tranquil-
izer that came from France; reserpine, which originated in India and
Switzerland; and meprobamate, a muscle relaxant that also allays
anxieties, which hailed from New Jersey.

When most people speak of sedatives, antianxiety medicines,
and antidepressants, they call them all tranquilizers, although the
three separate names show that the substances do not fit into one
category. The antipsychotic drugs, which control severe manias,
should be the only ones called tranquilizers, but the term *neuroleptics*
has become an accepted substitute for *antipsychotics*. Neuroleptics such
as chlorpromazine do not alter consciousness or thinking, do not
decrease initiative, but do make the patient less responsive to things
that might harm the sense of self-worth or psyche. This definition
overlooks the fact that most neuroleptics make patients at least
somewhat sleepy, but that is a small price to pay for the social adjust-
ment they bring and for smoothing out unwanted excitement and
hostility as well as severe withdrawal.

The antianxiety agents (such as diazepam and meprobamate), also
called minor tranquilizers, depress overexcitability of certain nerve paths.
So do the sedative-hypnotics, although they act at different locations in
the brain. Sedative-hypnotics are used primarily by physicians to induce
and maintain curative sleep. These medicines share this property with
several antihistamines that have a sleep-inducing side effect so strong that
they have been chosen as ingredients for several over-the-counter sleep-
producing medicines. Users are not usually told that a hangover usually
awaits them the next day and that they may become uncooperative and
resentful until the drug has been eliminated.

Antidepressant drugs stimulate the central nervous system's
parts that control emotion and mood. They do this by preventing the
absorption of certain chemicals (biogenic amines) by nerve endings,
thereby freeing them for special biochemical uses. The antidepressants
have also been called *thymoleptics*, and some of them, *psychic energizers*.
Another drug—this time an inorganic one—is useful for mental disease. Lithium salts are used to blunt the manic phase of manic-depressive states (*biphasic depression*).

ANTIPSYCHOTIC AGENTS

The history of these and all the other important psychopharmacological drugs is an account of accidental observations, with only marginal intrusions of logic. A fancy name for this kind of development is serendipity. The only exception is reserpine, the main alkaloid of *Rauwolfia serpentina*, a shrub that grows in India and around the Pacific basin. Its pink and white blossoms decorate the southern slopes of the Himalayas. The bush was called *pagla-ka-dawa* (insanity herb) by the Sherpas; its Sanskrit name in Ayuvedic medicine was *sarpagandha*. The French botanist Charles Plumier (1646–1704) in 1703 called it *Rauwolfia* in honor of a 16th-century Augsburg botanist, Leonhard Rauwolf (1535–1596), and *serpentina* because its roots are snakelike. Historians have found no evidence that Rauwolf ever saw the plant or even heard of it.

As was the custom with medicinal plants of old, rauwolfia was used to treat many different conditions—in alphabetical order, cataract, cholera, epilepsy, insanity, insomnia (which may have been due to high blood pressure), and snakebite (because of the shape of its roots!). Modern Indian chemists extracted the plant and found that the extract was active in the treatment of high blood pressure (*hypertension*) and psychoses. At first nobody paid any attention to the publication of these findings in an obscure Indian journal, but 25 years later, in 1949, one of the Indian authors (R. J. Vakil) described his results in a British journal, and that started the ball rolling. Reserpine was isolated and characterized in the Ciba laboratories in Basel, and the complicated alkaloid was synthesized by Robert Woodward (1917–1979) of Harvard University, probably the greatest organic chemist of this century.

Although reserpine has lost ground in psychiatry because of the depression that accompanies its use, it is still used to treat some forms of high blood pressure. It acts by releasing several biogenic amines from nerve endings, thereby making these neurohormones available for their mood-regulating tasks.

Chlorpromazine was developed through a series of studies that started with some antiallergic (*antihistaminic*) drugs derived from an organic chemical called phenothiazine. As mentioned before, some antihistamines make the user sleepy. This depression of the central nervous system is a prominent side effect and not their principal activity. French chemists led by Paul Charpentier tried to separate the allergy control from the depressive effect by molecular modification. Most classical antihistamines contain a chain of two carbon atoms connecting two nitrogen atoms, and this chain was made longer by synthesis. The resulting chemical had an amazing variety of biological activities, which made it hard to classify and to study for one activity at a time. In fact, the French called it Largactil for its "large actions." One clinician noted that it lowered the body temperature, and he used the compound to cool patients before heart surgery by inducing artificial hibernation. Cooling the body in cold, wet sheets was an old method of treating raging insane patients. The new drug, chlorpromazine, was therefore given to these psychotic patients by the psychiatrist J. Delay in Paris, with excellent results. In this country, chlorpromazine was first given in a gingerly manner as an antiemetic to counteract vomiting but was finally tested against psychoses under the trade name of Thorazine in 1954. It soon became the standard drug for treating psychotic patients.

It is not surprising that a drug with so many activities has a number of unwanted clinical side effects. The most objectionable of these are the extrapyramidal symptoms that some patients develop after prolonged treatment. These symptoms take the form of tremors similar to those in Parkinson's disease, the shaking palsy. This problem became the scientific driving force for trying to improve chlorpromazine by molecular modification. An estimated 10,000 similar compounds have been prepared and tested, a monumental worldwide effort that produced a dozen or so drugs now in clinical use. The most acceptable of these are thioridazine, clozapine, and risperidone, which cause few side effects.

An equal motivating force to modify chlorpromazine was the desire of almost every pharmaceutical firm to secure a share of the market commanded by this drug, as shown by its spectacular curative and economic results in hospitals for mental patients. Within three years after the introduction of chlorpromazine, most of these inadequately staffed, overcrowded hospitals were emptied because psychotic patients could be released and returned to their homes and useful occupations. The savings in public and private funds were enormous. Construction of new mental health facilities could be halted. The state of New York alone saved approximately two billion dollars over a seven-year period. The gain in human happiness and the joy of reunited families cannot even be estimated.

The effects of antipsychotic drugs—and also of antidepressant and antianxiety agents—changed the public's attitude toward mental disease. Regarded with superstitious fear, distrust, despair, and rejection since ancient times, psychoses and depressions are now generally recognized as functional diseases, not unlike diabetes, goiter, and other organic sicknesses that can be corrected. An attitude of kindness and understanding has replaced the horror of centuries of ignorance in clinical and personal relations with mental patients.

The feverishly competitive research on phenothiazines in the 1950s and 1960s not only resulted in new drugs but also led to the invention of biological tests that provided a better comparison with human psychotic states. That research resulted in finding behavioral tests for laboratory animals that had been given test drugs. The most common of these tests are based on what is called conditioned avoidance behavior and "learned" responses to the drug. The drugs being tested are meant to return the patient to normal behavior. Typically, an animal is placed in a cage with an electrically wired bottom. There is a wooden pole or a rope overhead. The animal is taught to avoid an electric shock to its feet by climbing the pole or the rope. Under the influence of an effective test drug, the escape by climbing will be delayed and the effect of a given voltage diminished. Or an animal will first be rewarded by food or drink for pressing a lever, then, when treated with a useful test drug, this learned response will be slowed. A whole battery of behavioral tests is usually employed to study the various actions of a test compound.

With such methods, it became possible to screen thousands of diverse chemicals for chlorpromazinelike reactions. From these tests emerged the butyrophenone class of antipsychotics, of which about 10 are used abroad. Only one—haloperidol—has been approved in the United States. Haloperidol was developed by the Belgian medicinal chemist and pharmacologist Paul A. J. Janssen through a series of chemical comparisons to other drugs, but not to chlorpromazine.

The starting point of Janssen's researches was meperidine, a morphinelike pain reliever known since the 1930s and studied in Germany before World War II. One American proprietary name is Demerol. Janssen made a large number of substances similar to meperidine and screened them for several activities, among them the then-fashionable test for conditioned avoidance. (Scientists follow fashions like the rest of the world.) The butyrophenones were vague and rather farfetched variations of meperidine, really not related well to that drug. They proved to be more potent than chlorpromazine but, unfortunately, shared with chlorpromazine the objectionable Parkinson-type reactions. Haloperidol is preferred to phenothiazine drugs by some psychiatrists but not by others.

On the way to the butyrophenones, about 4000 intermediates and analogues were prepared and screened, among them diphenoxylate, which turned out to be a useful cough suppressant but worked especially well for diarrhea by relaxing the intestines. It has become a standby for travelers visiting foreign countries.

The accepted mechanism of action of antipsychotic agents is that they block brain receptors for dopamine and serotonin. Antipsychotic agents are used in psychiatric practice to treat insanity and psychoses. Neuroses are not usually treated with these drugs but rather with antianxiety agents or sedatives, or by psychoanalysis.

ANTIDEPRESSANTS

Antidepressants relieve the deep depression arising from a disturbed concentration of essential chemicals (biogenic amines) in brain tissues (such as the *corpus coeruleum* and the *substantia nigra*). Depression caused by grief, personal loss, and similar temporary conditions do not respond to typical antidepressants. They can be relieved by antianxiety drugs or by newer agents such as fluoxetine (Prozac) which inhibits serotonin (5-HT-2) receptors and thereby regulates brain levels of serotonin.

To understand this difference, one must know a little of the physiological origin of internal (*endogenous*) depressions. "Normal" moods and emotions are maintained by the reactions of certain neurotransmitters at brain receptors that work much as previously explained in conducting currents from one nerve cell to the next. Of these neurotransmitters, three biogenic amines—dopamine, norepinephrine, and serotonin—are most important. A loss of these amines at their brain receptors comes about in two ways. One is the return of the amines into storage granules in nerve endings. These storage granules are microscopic reservoirs that serve as amine storage bins. The biogenic amines are pumped back into these reservoirs with the aid of neurotransmitter transporters. The other way is destruction of biogenic amines by enzymes, primarily by monoamine oxidases, or by enzymes that alter the neurohormones by a chemical reaction called methylation.

The first antidepressants owed their discovery to the observation that one of the drugs for tuberculosis introduced in the 1950s caused unwanted, serious mental disturbances as a side effect. The Swiss pharmacologist P. Zeller became interested in this troublesome property and soon found by trial and error that the antituberculosis drug iproniazid blocked the action of the monoamine oxidase enzyme. Iproniazid was then tried clinically in severely depressed patients, and it elevated their moods. Although it soon had to be withdrawn because it damaged the liver, its discovery pointed the way to the use of monoamine oxidase inhibitors as antidepressants. By blocking monoamine oxidase, such compounds prevent the destruction of the neurotransmitters needed to maintain serenity.

A number of analogues and more remote relatives of iproniazid were tested because the easiest way to achieve something is to do what someone else has done. Three of these analogues, phenelzine, isocarboxazid, and nialamide, have survived as useful antidepressants. Screening of other assorted chemicals revealed two drugs for the same purpose. One was tranylcypromine, a novel relative of amphetamine; the other was pargyline. Pargyline also lowers the blood pressure, and the manufacturer of this drug preferred to introduce it for this lucrative use.

Tranylcypromine is a potent antidepressant. This and many other monoamine oxidase blockers cause erratic changes in blood pressure if the patient is on an uncontrolled diet. The drugs prevent the destruction of tyramine, a widely distributed dietary substance that drives up the blood pressure. It occurs in cheeses such as cheddar, Stilton, and Camembert and in beer, bananas, canned figs, and red wines. Patients using monoamine oxidase blockers are warned not to consume such foods. Human nature being what it is, such a prohibition can be completely enforced only in a hospital.

Pargyline and tranylcypromine are among the first drugs whose mode of action is fully understood at the enzyme level. Both form known compounds with monoamine oxidase that change the enzyme so that it can no longer break down (oxidize) the neurotransmitter. In a way, the enzyme commits suicide by these chemical reactions. The two drugs are therefore spoken of as suicide-enzyme inhibitors.

Recently chemists have learned of another way that monoamine oxidase blockers affect the fate of neurotransmitters. They slow down the return of these biogenic amines into storage spaces in nerve endings, although not as efficiently as imipramine and other tricyclic antidepressants, but they increase this action when given together with these drugs.

The early tricyclic antidepressants had three rings of atoms in their structure, hence the name. They were conceived by Richard Kuhn (1900–1967) in Switzerland as variants of certain tranquilizers and were tested as such, but they behaved unexpectedly as antidepressants. The two standard-bearers of this class are imipramine and amitriptyline. As mentioned above, they maintain the neurotransmitters at nerve endings by preventing those biogenic amines from returning to their reservoirs.

Imipramine and amitriptyline are metabolized in the body to compounds that contain one carbon atom less (desipramine and nortriptyline, respectively). These metabolic products, now manufactured synthetically, are still effective antidepressants. Imipramine has also been found to prevent panic attacks and to alleviate bed-wetting.

Clinical depression is encountered in 15 million Americans at some point in their lives. Antidepressant therapy has done much to relieve the symptoms of such episodes which, in severe cases, may lead to suicide.

ANTIANXIETY AGENTS

Anxiety is an uneasy state of emotions akin to fear. It is a response to various disquieting environmental stimuli and is best treated by removal of such worrisome factors. If that cannot be done, an anti-anxiety drug may be prescribed. In fact, anxiety about other symptoms usually drives patients to see a physician, and such drugs are often the first aid in setting the patient on the road to recovery.

It is hard to devise tests for animals that correspond to human anxiety. Every time a laboratory animal is picked up and handled, it is anxious because it regards the researcher as a danger. But the symptoms of animal anxiety cannot be judged and compared because animals cannot verbalize their fears and feelings. Therefore, other actions of drugs, such as muscle relaxation, increased or slowed heartbeats and breathing, and reduction of twitching responses to unpleasant stimuli, have become part of a battery of tests designed to screen anti-anxiety compounds.

Conventional wisdom claims that one cannot be anxious and relaxed at the same time. Fortunately, it is easy to measure muscle relaxation, and therefore antianxiety agents have their origin in drugs that relax skeletal muscles. Frank M. Berger pioneered in screening compounds for this quality, and he selected a simple drug, mephenesin, as his prototype. More than 1200 compounds were tested before the best, meprobamate, was chosen for in-depth clinical study. Synthesized by B. J. Ludwig, this drug has become a widely prescribed antianxiety pill.

Berger named meprobamate Miltown for the New Jersey town he lived in at that time. He was president of Carter Laboratories, which had previously profited from the sale of Carter's Little Liver pills. When meprobamate came along, the company had no manufacturing facilities and turned to a larger firm, Wyeth Laboratories, for a cooperative agreement. Wyeth agreed to make meprobamate for Carter on condition that they could market part of it under their own trade name, Equanil. There are stories of patients who had taken Miltown but complained that it did not help their anxieties. The physician then changed the prescription to Equanil, and the grateful patients reported glowing success with the "new" drug. They did not know that both medicines had been made in the same batch and divided by the two companies. This example illustrates one of the difficulties of treating psychosomatic disorders, in which the mind upsets the body.

The commercial success of meprobamate stimulated many other pharmaceutical companies to search for similar drugs. There were not many leads, and therefore all kinds of compounds were screened in the battery of tests that Berger had published. The road to the benzodiazepines, which were destined to become the standard-bearers among antianxiety agents, began in these screening experiments.

Leo H. Sternbach (1908–   ) worked at a Polish university after 1930 on some organic chemicals that were thought to contain a ring of carbon, nitrogen, and oxygen atoms, but the methods for proving this structure were inadequate then. The work was stopped, and later Sternbach emigrated from Poland and found a job as a research chemist with the large New Jersey pharmaceutical firm Hoffmann-LaRoche, Inc. After a while, he was encouraged to spend part of his time on any subject he liked, so he picked up his unfinished old studies. With the modern instruments available, he found that the earlier chemical analysis was wrong, and that the compounds had a different structure. A few were tested but proved uninteresting, and the project was abandoned. In May 1957, the laboratory was cleaned, and a leftover vial of one untested compound was taken off the shelf. Because it had not yet been screened, it was sent to Lowell O. Randall, who tested it for meprobamatelike relaxing activity. The compound was so successful that clinical trials seemed indicated.

The drug, called chlordiazepoxide (which later became Librium) still had an adventurous fate ahead. Today, the chemical structure of a compound is ordinarily well known before biological testing is begun. If it turns out to be biologically interesting, chemists can make variants for patent claims and can also explore whether any of them might be more powerful or more suitable. Here, however, was a drug destined to become an excellent one for reducing anxiety even though its chemistry was not understood. A strange and unexpected chemical reaction had occurred during its synthesis, and it took months before the reaction could be explained. Only then could molecular modification begin.

These modifications soon furnished a whole string of "-azepams" that were improvements in several ways over Librium. Most important was a simpler one called diazepam, which, under the proprietary name of Valium, became the most widely prescribed of all medications and remained at the top of this bestseller list for several years. It is a good antianxiety agent.

Among other variants are chloro- and fluoro-derivatives called lorazepam and flurazepam, respectively. They are prescribed for insomnia, and they also help to overcome jet lag. They cannot be abused to commit suicide even in large doses.

SEDATIVE-HYPNOTICS

The word *hypnotic* comes from the name of the Greek god of sleep, Hypnos. Before people or animals go to sleep, they get drowsy, and before that they become calm. There is a steady transition from wakefulness to sedation to the various stages of sleep. Different drugs lead to intermediate stages or all the way beyond REM (rapid eye movement) sleep. Scopolamine produces twilight sleep; phenobarbital, deep sleep.

Millions of people suffer from insomnia for a variety of environmental or internal reasons. Millions of others think they cannot sleep, although in reality they are asleep much of the time most nights. Many people seek sleep through the aid of drugs at least occasionally. The frequent advertisements of sleep aids on television bear witness to the need for and profitability of such drugs. Ten or twenty years ago most of the over-the-counter hypnotics contained scopolamine; today the active ingredient is usually diphenhydramine, an antihistamine that blocks the H-1 receptor and has a strong component of antispasmodic and sedative (*anticholinergic*) activity.

The best-known older hypnotics were the barbiturate drugs. They are now mostly of historic significance. They were excellent agents to combat insomnia but have been abused widely to produce nirvanalike mental states. In overdoses they also provided suicide aids to deeply depressed persons. The manufacturers of these barbiturates did not wish to have their good names involved in illegal affairs and stopped making these sedative-hypnotics. However, some of them are valuable for other conditions. Phenobarbital, first introduced in 1912, is a powerful hypnotic, but its main use is as an effective broad-spectrum anticonvulsant in several types of epilepsy. Some sulfur-containing barbiturates, such as thiopental, produce general surgical anesthesia when injected into a vein. Their short action is due to the ease with which they are metabolized and eliminated.

Many persons who have slept well after taking one of the common hypnotics sleep almost as well a second night. The chemical reason for this is that these drugs are soluble in fats and other fatlike lipids. After acting on the brain for a while, they are swept into the blood and stored in the body fat. Later, the circulating blood moves them back to the brain, where they resume their sleep-producing activity the next evening. This cycle ends only when all the drug has been broken down and eliminated.

ETHYL ALCOHOL

Ethyl alcohol is probably the oldest drug humans have used. Prepared by fermentation of starch from grains (such as corn, barley, and rice) or grapes under the catalytic influence of yeast enzymes, the resulting dilute aqueous solutions of ethanol are concentrated by fractional distillation. The fraction that boils at 78–79 °C is pure ethanol; it can be diluted to 41–49% and flavored as whiskey, vodka, gin, or liqueurs. Wine contains up to 10% and beer up to 4% ethanol.

Alcohol first stimulates, then depresses the central nervous system, and finally produces toxic symptoms such as nausea, vomiting, incoordination, and stupor. It can lead to addiction (alcoholism). Efforts to withdraw ethanol from alcoholic patients require both drugs and psychotherapy. Among the drugs is daidzein, the active ingredient of kudzu, an imported nuisance weed that often chokes trees in forests in the southern United States. It also occurs in soybean meal, red clover, and the mold *Micromonospora halophytica*. It has been synthesized. Another drug for alcoholism is disulfiram; it causes intense vasodilation, tachycardia, and other toxicity symptoms. It has a foul odor, which persuades alcoholics to abstain from drinking alcohol.

BRAIN STIMULANTS

The neurotransmitters such as dopamine and serotonin are essential to maintain the mood and normal emotions. They share this "tonic" property with several similar compounds, both natural and synthetic. As previously mentioned, the alkaloid ephedrine, from the desert shrub, *Ephedra vulgaris*, stimulates the brain's cerebral cortex. This plant's activity has been known in China for thousands of years. When the modern pharmacologist K. K. Chen demonstrated this property to a class of medical students, he dosed a dog with ephedrine, and the class watched the increasing restlessness of the animal in a jiggle cage. After the students filed out, Chen, as an afterthought, measured the dog's blood pressure and found it above normal. It then occurred to him that ephedrine's chemical structure is similar to those of the biogenic amines that raise blood pressure by constricting blood vessels.

Several substances with chemistry similar to that of ephedrine were known and could be expected to give similar results. One of them, even simpler than ephedrine, was synthesized in Romania in 1887 but had been forgotten. After Gordon A. Alles prepared it again and tested it, he found that it was a good vasoconstrictor and a strong brain stimulant. The compound, named amphetamine, opened nasal passages that had been narrowed by swollen blood vessels. It also became the standard drug for people who had to keep their attention on their work, such as driving trucks and piloting aircraft, and who had to overcome tiredness. Amphetamine is a mixture of equal parts of two enantiomers. Only one of them, dextroamphetamine, produces the drug's desired effects. Because dextroamphetamine also overcomes emotional sluggishness, it has been widely used as a way to reduce overweight. People gain weight by overeating, often because of unhappy, lonely, or unsatisfying experiences. By overcoming such depressed feelings, dextroamphetamine helps them to stop stuffing themselves and thereby to lose unwanted pounds. A similar medicine, phenylethanolamine, is used for the same purpose.

A derivative of amphetamine that contains an extra CH_2 group in the amine portion is called methamphetamine; it is about as active as amphetamine and serves primarily as a brain stimulant. After the defeat of Japan in 1945, young Japanese people had to take whatever jobs they could find, both during the day and at night. Many workers turned to methamphetamine to overcome sleep while on duty because the drug was freely available without prescription. Soon many instances of frightening psychoses were traced to the use of methamphetamine, and prolonged use of amphetamine also caused such problems.

From a psychiatrist's point of view, these drug-induced psychoses became opportunities to study abnormal behavior under controlled conditions because these psychoses closely resembled several human mental diseases. It is different and more frightening when deluded people purposely take amphetamine to become psychotic. This problem of sociopathic drug abuse has not yet been brought under control.

Nor is it yet possible to devise a drug that stimulates the cortex of the brain without inviting abuse. Boredom starts this abuse in those who mistakenly believe that removal of some uncomfortable symptoms will improve underlying conditions. The same story has been repeated ominously with cocaine, which also stimulates the central nervous system. Authorities can try to prevent illegal synthesis of drugs like amphetamine that require some skill to make, but it is almost impossible to control the simple extraction in water of coca leaves in poor South American countries, where the economy depends on immense profits from smuggling tons of cocaine into developed countries with a free currency. The fact that their cocaine is never really pure and usually the volume is increased by substances that may be harmful does not bother these unlawful operators or those that distribute the cocaine locally.

LITHIUM

Lithium salts occur naturally in seawater and in many mineral springs. Lithium urate, the lithium salt of uric acid, is, at its higher pH, more soluble in water than is uric acid, which is a normal minor end product of the metabolism of nitrogen compounds. Patients suffering from gouty arthritis produce more uric acid than healthy individuals, and the insoluble excess is deposited as crystals in their joints. Such patients benefit from drinking lithium salt solutions because they can then excrete lithium urate and thereby decrease their gouty deposits. For this reason, affluent gouty patients used to drink lithium-containing water at elegant spas, although sometimes the vacation itself may have contributed as much to their recovery as the lithium diet. In any event, lithium therapy has had a long history, so anybody inventing a new use for lithium could bypass the time-consuming study required by the FDA's regulations.

The Australian psychiatrist John F. J. Cade, who studied manic-depressive illness in 1948, reasoned that abnormally high concentrations of some product in the body might cause this condition. He also felt that too low concentration of such a substance might be the origin of melancholia, the old name for severe depression. He analyzed the urine of manic-depressive patients but could not find an unusual excretion product. Nevertheless, when the patients' urine was injected into guinea pigs, the animals died from fatal convulsions at one third of the deadly dose of normal human urine. So something in the urine from manic-depressives must have been more toxic to the animals. In a roundabout way this toxicity was attributed to uric acid. The investigators intended to prove this idea by adding uric acid to normal urine, which should have made it more toxic. But uric acid was too insoluble to test, and therefore the more soluble lithium salt was used. To their surprise, lithium urate decreased the toxicity of urine. Then when lithium carbonate was studied alone, without urine, the animals became tranquilized. This finding was an open invitation to try lithium salts for mental patients. Schizophrenic and chronic psychotic depressive patients did not respond, but manic patients improved so much that they could be discharged from the hospital.

What does such a simple alkali ion as lithium do to suppress manic episodes? Recent studies have shown that the amounts of some fundamental hormones of metabolism such as cyclic adenosine monophosphate (cAMP) and inositol phosphate change under the influence of a lithium ion. These changes may account for lithium's power.

DRUGS THAT EVOKE PSYCHOSES

When early humans tried out every kind of food they could find in fields and forests, they encountered mushrooms and other plants that made them feel peculiar. Later, brews stewed from such vegetation produced similar feelings. This reaction happened in Central and South America, in India and in Siberia, and probably all over the Earth. In this way, primitive people learned about plants that would raise their moods to unexpected heights or throw them into depths of fear and despair that they believed was punishment for toying with the secrets of nature. These early "scientists" experienced strange visions that they had not experienced before and glimpsed unknown insights. In the language of today's drug abusers, they had good trips and bad trips. They searched persistently to find such plants among the inexhaustible variety of the vegetable kingdom.

Priests and witch doctors of old gathered these plants for their own purposes. They used them themselves and gave them to their followers to induce detachment but also religious frenzy or abject fear of their deities. The plants and their decoctions were also used in celebrations, to cause stupor or drunkenness, warlike excitement, and indifference to danger. The drugs contained in these plants are sometimes called *psychedelic*, meaning mind-manifesting. Because some of the effects they produce resemble serious mental conditions or psychoses, the name *psychotomimetics* is also used for the group. The drugs change perception and mood, disturb the autonomic nervous system that controls normal mood, and in high doses, often cause hallucinations. Obviously physicians want to avoid these reactions, which are

regarded as unwanted and feared side effects. Such drugs are therefore not used to treat patients.

Natural products from the plant kingdom—and a few animal
sources such as toad skins—are not the only sources of mind-tampering
drugs. Many synthetic compounds, including heroin and LSD, behave in
the same way. These compounds have given drugs such a bad name that
many people overlook the value of most drugs for medicinal purposes.
This negative reaction comes from the unhappiness, misery, crime, and
death that have resulted when psychotomimetics are abused. Millions of
people all over the world have fallen into this trap, from which addiction
and personality weakness often bar their escape.

Other drugs that have clinical uses but also have pronounced
depressive or stimulatory side effects are excluded from this discussion because, although they are abused by unbalanced persons, their
medicinal value outweighs these unfortunate occurrences. Among
these drugs are morphine, atropine, scopolamine, their derivatives,
and other potent analgesics and sleep inducers. All these medicines
are useful, with side effects that are not too pronounced when used
briefly. Prolonged use, however, may lead to dependence.

One of the oldest psychotomimetic drugs was soma, a concoction of unknown botanical origin used in India thousands of years
ago. Sanskrit manuscripts say that it made one feel like a god. Similarly, ancient religious oracles and cults like the ones at Delphi and
Eleusis in Greece drugged priestesses to visionary incoherence.

The hemp plant, *Cannabis sativa*, dates back at least to the time
before Herodotus (circa 484–425 B.C.), who wrote that the Scythians
on the Caspian Sea used it for self-intoxication. The crusaders encountered hashish in the Middle East made from cannabis. It was used by
terrorists sent on missions of assassination; the name *hashish* means
assassinate. This drug spread to high society in Europe as a means of
escaping boredom, and its sinister threat culminated in this country
around 1955, when millions of young men and women began to
smoke the resin of the plant. A more diluted version is called marijuana. It is not a narcotic-hallucinatory drug, but it depresses the brain,
causing intellectual indifference, ineffectiveness, listlessness, and loss
of productivity. Prolonged use can destroy positive personality traits.
The ease of growing and harvesting hemp and the ability of South
American and Asian smugglers to export marijuana to North America
and Europe have made control of this drug virtually impossible.

The mind-affecting constituents of cannabis are the tetrahydro-cannabinols. These compounds produce a sedation of the central nervous system at low doses. They have helped some asthmatics and have quelled nausea for some people taking anticancer drugs. Fiber from hemp is used in making certain kinds of rope.

Central America has supplied more mind-upsetting plants than any other part of the world, probably because Mexican and Peruvian jungles have been more accessible than African or Malayan tropical regions until recent times. Also the Aztecs, Otomacs, and South American Indians as well as the Native Americans of New Mexico and Arizona have kept up religious cults and ceremonies in which plants play a role despite centuries of attempts to convert them to various Christian sects. This type of social culture has helped botanists to locate sources of such plants.

One of the first magic drugs discovered in Mexico and the American Southwest was *Lophophora williamsii*, also called *Anhalonium lewinii*. The above-ground parts of this spherical cactus (mescal buttons) contain a number of alkaloids related to dopamine. The Indians prepare a drink from the cactus called *peyote* or *pyotl*. The drink itself, as well as mescaline and other alkaloids isolated from the cactus, have powerful mental effects and produce vivid color visions. Mescaline is a totally disorienting drug that causes long-lasting and frightening psychotic episodes. Physicians can bring mescaline users under control slowly with tranquilizing (*neuroleptic*) drugs. Mescaline and many of its chemical cousins have been repeatedly synthesized and tested.

Another plant still being used by several Mexican tribes is ololiuqui, an herb with long white flowers and round seeds. Botanically it is called *Rivea corymbosa*, but it has a number of common names, including snake plant and herb of the Virgin. The crushed seeds, or an alcoholic beverage prepared from them, induce delirium, visions, satanic hallucinations, or a narcotic type of sleep not unlike twilight sleep. The psychotomimetic substances in this plant are ergot alkaloids. Such compounds had been previously isolated from ergot, a mold growing on rye. Rye bread contaminated with ergot caused havoc in the Middle Ages in Europe, ranging from epidemics of abortions caused by uterine contractions to hallucinations called Saint Vitus' dance (chorea). Albert Hofmann, a Swiss biochemist, cleared up the chemistry of the new and different alkaloids from ololiuqui.

A plant called *teonanacatl* ("sacred mushroom") was regarded as a god by Mexican Indians and as the devil by Christian missionaries after the conquest of Mexico by Hernán Cortés. This mushroom contains two compounds that affect the mind, psilocin and psilocybin. Chemically they are related to the neurohormone serotonin. Chemists have synthesized these fairly simple compounds, and psychiatrists have tried to use them as aids to psychoanalysis and psychotherapy. The ancient cult of the mushroom is based on colorful and unreal visions that occur as the mushroom is eaten.

Indians of the Western Amazon basin brewed a magic potion they called *ayahuasca, caapi,* or *yajé*. It is made from plants of the *Banisteriopsis* genera, which contain toxic alkaloids. The same alkaloids occur in a plant (*Peganum harmala*) that grows in northern steppes; its seeds were used in Arabian medicine to treat worms and to promote sweating. The harmala alkaloids are powerful blockers of the enzyme monoamine oxidase, but probably are not responsible for the mental effects of the parent plants.

In the Orinoco basin of Venezuela, the seeds of some members of the pea family (*Piptadenia*) are ground, mixed with lime, and used widely like tobacco snuff under the name of *yopo*. Men and boys blow the snuff into each other's nostrils through a forked tube made of chicken bones. Indian witch doctors use it to inspire prophecy and clairvoyance. *Piptadenia* contains several chemically simple indole alkaloids, among them bufotenine, which is also found in the skin of poisonous toads. The most dangerous of these substances is dimethyltryptamine, which is highly psychotomimetic. Because it is related to tryptamine, a normal biogenic amine of brain metabolism and can be made biologically from this compound, some people think that schizophrenia may result from the presence of dimethyltryptamine as a faulty metabolic product in humans.

LSD, or LSD-25, is chemically named lysergic acid diethylamide. It is the most powerful and most specific psychotomimetic known. It is a synthetic compound, made by laboratory procedures, and—alas—has been at one time or another one of the most widely abused mind-altering drugs. Other amides of lysergic acid occur in the old Mexican magic plant ololiuqui and in ergot. LSD-25 was the result of standardized practice in medicinal chemistry.

The ergot alkaloids are complex derivatives of lysergic acid that have, among other properties, a uterus-contracting (*oxytocic*) effect that can be of use in obstetrics. Albert Hofmann in Basel, Switzerland, was given the task of modifying the structure of these compounds and testing their "cousins" for probable biological uses. He made many derivatives, one after another. The diethylamide was number 25. It was tested and, as expected, contracted the uterus in laboratory animals. It also strongly excited them.

At that point, Hofmann became sick, dizzy, and restless in the laboratory, so he went home, lay down, and had what we would now call an LSD trip. It lasted for two hours, and he tried to figure out what had caused it. The last thing he had done was to prepare LSD-25; therefore, he suspected that he might have inhaled a bit of this crystalline powder. Three days later he tried to confirm his suspicion by taking one quarter of a milligram of LSD by mouth. He barely got home after that and had a very "bad trip." His family called a doctor, who stayed with him during the crisis. This experience is not hard to explain today: the effective dose of LSD in humans is less than 0.05 milligram, but poor Hofmann took five times that much. LSD is 5,000–10,000 times more active than psilocybin from mushrooms. Its effects and dosages were confirmed in volunteers, including graduate students of the University of Basel, who still remember their experiences with a shudder.

Since the first days of the research on the classical mind-altering drugs, many other substances have been found that cause profound psychic changes. Almost all of them have side effects on the autonomic nervous system, and therefore the mental effects are accompanied by changes in heart rate, intestinal irregularities, difficulties in breathing, and abrupt changes in blood pressure. These drugs make people very sick who set out only to expand their minds without understanding or believing the horrors and pains that result from taking these drugs of ill repute.

The newer horror drugs have chemical names that the users, who are often school dropouts, cannot pronounce, so they give them an assortment of alphabet-soup initials, which are also used to refer to them in the media. They are compounds made experimentally that have been smuggled out of research laboratories. In one case a drug had been tested as a general anesthetic before it was appropriated by abusers.

Some of these newer agents have been called synthetic heroin or designer drugs, although they are chemically unrelated to heroin and have not been designed. Two of them were side products in the manufacture of the analgesic meperidine. Others are synthetic compounds tried out by addicts in the hope that they might give them a new mental high. The most dangerous of these materials are 3-methylfentanyl and MDMA. Both drugs produce dangerous damage to the general health of the users and cause heroinlike addiction at unbelievably low doses.

These tragic experiences with the abuse of drugs raise questions of what can be done to curb abuse and whether it should be punished or allowed to disappear like other manifestations of a temporary culture. In considering whether drug abuse should be controlled and prevented, a detached scientific point of view requires judging each situation on its own merits or demerits rather than lumping unlike problems together indiscriminately. The abuses of heroin, cocaine, and marijuana can serve as examples.

Heroin is an addictive narcotic; cocaine is an addictive stimulant; marijuana is a nonaddictive minor depressant of the central nervous system, with no actions that reach into the narcotic stage. Heroin causes an uncompromising craving for renewal of the narcotic dose; it is unreasonably expensive because its price is driven up by criminal suppliers and distributors. The addicted victims of cocaine present a picture different from that of heroin addicts. Cocaine addicts can quit only with difficulty, but without the dreaded withdrawal symptoms of the morphine-heroin type. Marijuana users can quit most easily, without any more physical discomfort than that felt by chronic tobacco smokers or coffee drinkers. At the bottom of the failure to stop using this drug is psychological weakness and lack of stamina—insufficient will to lead a drug-free life. Heroin withdrawal requires medical assistance; marijuana withdrawal requires a strong personality. Cocaine withdrawal falls between these two.

At present, cocaine presents the principal abuse problem in the industrialized world. For many years, the hydrochloride salt of cocaine had been used both for oral and intravenous administration. In the late 1980s, the free base of cocaine began to be sold under the name of crack. This crystalline material is insoluble in water, but can be inhaled by smoking, and acts more forcefully and dangerously than oral cocaine hydrochloride.

The second most dangerous drug of abuse is methamphetamine. It is synthesized readily and can be used in the form of water-soluble salts, or as the free base sold under the street name of ice. It causes paranoid psychoses and violent behavior, especially during the rapid and intense "high" that occurs immediately after smoking ice.

No legislation will prevent people from experimenting with drugs, not even in a police state. Exploring strange life situations, revolting against parental authority, and experimenting with sex and drugs has been going on for thousands of years. We happen to live near the crest of a wave of drug abuse, as has happened before in other times and places. The danger is that widespread drug abuse—apart from its criminal aspects—may lead to lassitude, social indifference, loss of initiative, and other factors that damage the integrity of a civilization. China before Mao had sunk into a state of lowered stamina, not only because of undernourishment of the population, but because of widespread use of opium and hashish and the ensuing impoverishment of the users and their families. The Roman Empire crumbled perhaps because of drug abuse. These examples should be a warning to Americans, who—because of their wealth and geographical nearness to drug-producing countries—are the easiest target for drug dealers. It will take superhuman wisdom guided by clinical psychologists and experts in public health to make a dent in drug dependence and abuse by individuals.

Unstable individuals will continue to seek escape from unhappy states of mind, whatever the cause of their condition. They know the legitimate use of sedatives and antianxiety agents. But do the clinical effects of these chemicals alter such states as insomnia, worry, fear, despair, and panic? These chemicals calm and straighten out disturbed mental conditions, and therefore the potential abusers try them out to alter their often self-imposed miseries. Larger and larger doses may be required to satisfy their craving for nirvana, and the drug users have less and less money to buy the drugs on the black market. So they undertake ill-planned robberies or even worse to gain possession of mind-altering agents. That action brings them in conflict with law enforcers, then they get prison terms, are released, and start all over again. Every drug that has a direct or indirect action on the central nervous system, whether stimulating or depressing within medical limits of application, has been tried and undoubtedly will be tried by these unfortunate seekers of happiness who do not have the fortitude to raise themselves by their bootstraps.

ANTICONVULSANTS

Many anticonvulsant or antiepileptic drugs are chemically similar to sleep-producing hypnotics. They also sedate the brain, although usually less so than the standard hypnotics. They control epilepsy by preventing or stopping the sudden bursts of electrical discharges between nerve cells in the brain that are characteristic of epileptic attacks and lead to convulsions. Some minor epilepsies produce only a dreamlike momentary unconsciousness (*petit mal*), whereas grand mal epilepsy results in generalized convulsions and loss of consciousness for an appreciable time. The explanation of epilepsies as faulty electrical discharges in the brain has not removed fear, but it has done away with persistent superstitions that accused epileptics of harboring evil spirits that had to be exorcised. Saints and witches of old were often afflicted by epilepsy and were persecuted as heretics or as possessed by the devil.

Animal tests for anticonvulsant activity are relatively simple and have served to stimulate research on antiepileptic drugs, the first of which, introduced clinically in 1911, was phenobarbital. Molecular modification led to several similar products (including primidone and mephobarbital), each of which has certain advantages and disadvantages. One disadvantage is their cost. Epileptic patients must protect themselves from convulsions for the rest of their lives, and an inexpensive drug such as phenobarbital may be preferred for this reason. A chemical analogue, phenytoin (formerly called diphenylhydantoin), followed phenobarbital 26 years later and is said to cause less sleepiness during the day. Each anticonvulsant counteracts a different array of electrical discharges, so the most valuable of these drugs are more suitable for epileptic children. In recent years, some chemically novel anticonvulsants have been discovered by screening; the most notable are valproic acid, a simple organic chemical, and felbamate. These compounds affect the metabolism of brain chemicals such as GABA, which may account for their effects.

DRUGS FOR PARKINSON'S AND ALZHEIMER'S DISEASES

In 1817, James Parkinson (1755–1824) described a disease, *paralysis agitans*, in his "Essay on the Shaking Palsy." This disease, which now bears his name, has a phase of tremors and abnormal movements and another one of rigidity. Parkinsonism is characterized, perhaps caused, by the almost complete lack of the neurohormone dopamine in certain brain regions (*corpus striatum* and *substantia nigra*). Parkinsonism is therefore a deficiency disease. Replacement of dopamine in these tissues might alleviate the symptoms of the disease. Getting dopamine into the brain is a problem, however.

The plastic, spongy material of the brain and the spinal cord consists mostly of lipids, which are water-insoluble substances similar to fats. The main difference between brain tissue and other tissues is that although permeated by watery channels, brain tissue is not easily accessible to acid or basic compounds. It behaves as if it were enclosed in a capsule, which is called the blood–brain barrier. Only neutral substances can get through this layer of minute blood vessels and lipids. Dopamine is basic (alkaline) and cannot pass the barrier, but the more neutral precursor from which it is made in body tissues can pass. This precursor is called levodopa. In the brain, but also in other tissues, levodopa loses carbon dioxide and becomes dopamine, which is needed to prevent parkinsonian tremors. Large doses of levodopa must be taken to get a small amount through the blood–brain barrier. This overdose leads to side effects. Most of the drug loses carbon dioxide through the action of an enzyme in the liver and kidney, the same enzyme that is needed in the brain to convert levodopa to dopamine in that organ. In the United States, the compound carbidopa is available for this purpose. It keeps levodopa from enzymatic destruction until the mixture reaches the blood–brain barrier. Only levodopa passes this hurdle, while carbidopa stays behind.

Symptomatic relief from the tremors and muscle spasms of the disease can be provided by natural and synthetic antispasmodic drugs and by certain antihistamines that also block acetylcholine. The antiviral agent amantadine also improves the tremors; some inhibitors of monoamine oxidase achieve the same result, especially deprenyl.

Similar tremors also occur as side effects when patients with mental illness are treated with antipsychotic drugs. These symptoms can be abolished by some of the anticholinergic agents.

Parkinsonism is only one of many degenerative diseases that may develop at any age but are most prevalent in older patients. Because drugs have cured many infections that used to kill the young, the world's population is statistically older than in the past. Therefore, more of these diseases of aging have moved to the center of medical interest.

The best way to avoid degeneration would be to stay physiologically young for a long time. Many ways have been suggested to achieve this, but none of them has remained unchallenged. By following some of these ideas and hunches, medicinal scientists have tested various chemicals for their ability to stabilize the metabolism and slow physiological deterioration. These chemicals include substances that counteract faulty synthesis of proteins by preventing errors in the genetic information that controls formation of protein in the body. One of these chemicals is the local anesthetic and heart drug procainamide. Several rejuvenating clinics in Romania and Switzerland specialize in injecting procaine into gullible individuals, promising them renewed youthful vigor. More to the point is the use of substances (vitamins D and E and carotene) that are known to prevent oxidation of certain unsaturated nutrients. The oxidized materials almost certainly contribute to malignancies and cell destruction.

Aging may also involve cross-linking of proteins (especially collagen). Cross-linking increases the size of fibers and causes local or generalized tissue toughening. This toughening occurs in Alzheimer's disease, a devastating progressive dementia that afflicts an estimated 4 million elderly people and causes 100,000 deaths annually in the United States alone. On autopsy, the brains of Alzheimer's patients show plaques, localized abnormal patches of tissue, often in the blood vessels of the brain. These plaques appear to be due to a mutant version of an amyloid ß-protein (AßP). This protein is relatively small, consisting of 39–43 amino acid residues, and is normally found in the brain. On alteration of two amino acids by mutation, however, it becomes six times as abundant and thus deposits many more amyloid plaques. No way has yet been discovered to prevent the mutation process that leads to the disease.

As in so many other diseases, peripheral processes that accompany Alzheimer's disease are being studied with the hope of reducing the dramatic loss of memory and other symptoms. There is a loss of the parasympathetic neurohormone, acetylcholine, in Alzheimer's disease, and several known drugs that prevent the decomposition of acetylcholine to choline and acetate have been tried with varying benefits.

Less than half a century ago, mental and most neurological diseases could not be treated by the medical profession. Psychoanalysis, as practiced by the Austrian psychiatrists Sigmund Freud (1856–1939) and Alfred Adler (1870–1937), and the Swiss psychologist Carl Gustav Jung (1875–1961), succeeded in part in dealing with neuroses and depressive disorders, but failed in dealing with schizophrenia and other psychoses. As neurohormones and neurotransmitters were discovered by biochemists and characterized and synthesized by organic chemists, the etiology of these biocatalysts in neurological and mental diseases began to be understood. These diseases are metabolic or deficiency diseases, not unlike other hormonal failures such as goiter, diabetes, and dwarfism, and not mysterious manifestations of superstitious or sexual aberrations. Probably faulty genes are responsible for some of these diseases, and dietary and environmental factors may play a role in many of these afflictions. This idea is supported by the isolation of psychotomimetic compounds from mushrooms and other dietary products used to effect self-induced psychoses.

As the population gets older because mortality from infectious and childhood diseases has decreased, more elderly individuals present diseases discussed in this chapter. The great advances in radiological and spectroscopic procedures, including ultrasound and magnetic resonance spectroscopy, have given us a clearer understanding of nervous transmission and its failures by degeneration of tissues in which the essential transmitters are biosynthesized. These studies have given neurology and psychiatry a new direction in which the role of biochemistry has steadily grown.

9

Drugs for the Relief from Pain

Pain is the most aggravating manifestation of disease, and ever since the dawn of recorded history, humans have tried to lessen, blunt, and avoid pain in all its forms. Psychological persuasion, diverting attention from painful stimuli, can go only so far in decreasing minor pain. The best way to stop pain is to remove its cause, whether through the healing of injured tissue, the cure of an infection, or surgical or radiological treatment of damaged organs. When radical treatment is undesirable, patients try to decrease aches and pains through the use of drugs. The old Greek name for pain was *algos*, and chemicals that counteract pain are called *analgesics* or *analgetics*.

OPIATES

As in most other types of medicines, the early analgesics were crude natural products, mostly botanical, or concoctions made from them by primitive methods of extraction. Opium and products of plants growing in temperate zones such as bark of willow trees were among the first to be used. Opium is made by cooking the juice of unripe oriental poppies to a dark tarry mass, which is powdered and can then be smoked in long pipes. It brings a feeling of well-being followed by forgetfulness and sleep and helps to wipe out pain. This feeling of

89

euphoria or well-being is even more pronounced from morphine, the extracted and purified main alkaloid of the oriental poppy. Morphine relieves severe pain from operations, accident, and battle wounds, but it also brings about forgetfulness and sleep.

Codeine also occurs in opium but is manufactured in one simple step from morphine. It is only one-tenth as analgesic as morphine, and it can overcome medium- to low-grade pain. It is less addictive than morphine, though. It also suppresses the cough reflex and is still incorporated in some cough medicines, although the less constipating synthetic drug dextromethorphan has largely replaced it for this purpose.

The most notorious of the many hundreds of derivatives of morphine is heroin, which is made from morphine in one step. It rapidly causes a dependence that is particularly hard to shake. The drug has to be injected to be active, and impure samples cause infections and other organic damage.

Two synthetic analogues of morphine are meperidine and methadone, which are both widely used to dampen pain. Methadone is also substituted for heroin to wean addicts slowly away from heroin. Both meperidine and methadone were designed by German industrial chemists to control spasms; their power to control pain was later discovered by screening. Another widely used and moderately powerful analgesic is propoxyphene.

How the brain perceives pain and how powerful analgesics work are questions that have intrigued generations of scientists. In the early 1970s, a British biochemist discovered several small and medium-sized proteinlike compounds that were named enkephalins from the Greek meaning "within the head." These substances are short-lived but strong painkillers when injected into laboratory animals. The enkephalins are made in body cells from larger peptides called endorphins (from *endo* meaning inside and (m)orphines) whenever the need for pain relief arises. They account for the fact that people often feel little pain at first from accidents or wounds. It appears that the action of potent analgesics is conveyed to the brain by these natural peptides. Because the enkephalins are also found in portions of rat spinal cords close to nerve tissues containing norepinephrine, a connection between these two types of neurotransmitters has been suggested. Attempts to modify the enkephalins for use in controlling human pain for any length of time have been unsuccessful.

Acupuncture and acupressure, the ancient oriental methods of relieving pain or preventing the onset of surgical pain by inserting needles or pressing at specific places, is believed by some to activate the endorphin–enkephalin system in the central nervous system and thereby to produce a profound analgesia. This theory has not been proven.

None of the strong painkillers can be obtained without a prescription. The many side effects, particularly the danger of addiction, require that they be used only when absolutely necessary to control severe pain. Many patients who suffer lesser pains can obtain mild analgesics in any drugstore without a prescription. If such pains are caused by rheumatism or arthritic inflammation, as is often the case, both the old and the modern antiinflammatory drugs give effective relief.

ANTIINFLAMMATORY AGENTS

Two broad classes of drugs control the hot, painful swelling of inflammation: steroidal and nonsteroidal. The steroidal agents are more effective, but they also have more side effects. Therefore, they are prescription drugs. Many recently discovered nonsteroidal substances that soothe inflammation are too new to be offered as over-the-counter medicines and also require prescriptions. The older, nonsteroidal medicines—led by aspirin and followed by acetaminophen, best known as Tylenol—are the active ingredients of almost all the "pain and fever" medicines sold in the United States. Ibuprofen, another effective substance for moderate pain, is now also available without prescription.

Aspirin was already an old compound when the German pharmacologist H. Dreser introduced it to reduce fever and moderate pain in 1899. Aspirin is derived from salicylic acid, which had been used for decades to treat pain and fever. Salicylic acid, however, was beset with side effects, especially a tendency to cause stomach ulcers. Chemists thought that a modification that would release salicylic acid after passage through the stomach would have fewer damaging side effects, and dozens of such variants were synthesized and tested. Aspirin has been the most durable survivor of them all. Today's annual world consumption of aspirin is 25 million kilograms (25,000 tons)!

One can buy the same dose of aspirin as tablets and as capsules at costs ranging from a few pennies to a lot more. What is the difference? Aspirin is made from salicylic acid and acetic anhydride. When one opens a bottle of cheap aspirin tablets and smells the stuff, the stinging, vinegary odor of acetic acid is often unmistakable. The smell means that the drug has not been purified adequately or that it has decomposed on the shelf. These impurities may irritate the stomach. Naturally, the more carefully recrystallized brands are more expensive, but in this instance economizing does not pay. The manufacturers of the better brands want the public to know this, and every day prominent actors promote these products on television. We pay for that advertising in addition to the cost of purification and preparation of loosely pressed, more readily absorbed forms. Even so, aspirin (acetylsalicylic acid) remains acidic, and no trick of chemistry or advertising can change that. One way to lessen damage to the stomach wall is to coat the pill with some jellylike basic substance such as aluminum hydroxide (alumina). This substance buffers the acidity of the drug and reduces the likelihood of ulceration.

Doctors have long known that rheumatic heart disease responds well to large doses of aspirin. Small doses have been credited with reducing the incidence and recurrence of heart attacks. A newly recognized drawback of aspirin is that it may precipitate a severe disturbance (Reye's syndrome) in a very small proportion of young children. Aspirin is an organic acid. It has no monopoly on relieving rheumatic and arthritic inflammation. More than 100 other organic acids, among them ibuprofen, indomethacin, and sulindac, act in a similar way. These newer drugs have some advantages over aspirin. Some of them reduce inflammation better, and some are better for the relief of medium-grade pain.

The action of powerful opiumlike analgesics has been attributed to the intervention of the enkephalins, but the antiinflammatory drugs do not work that way. Seventy-two years after the introduction of aspirin, the British pharmacologist John R. Vane observed in 1970 that aspirin interferes with the biosynthesis of some of the prostaglandins within the body. These active chemicals were first isolated from the prostate gland's fluids (hence their name) by the Swedish biochemist Ulysses S. von Euler in 1935.

There are many natural prostaglandins, and now there are many hundreds of molecular modifications conceived in the laboratory that have diverse biological properties. One prostaglandin, PGE 2, increases the awareness of pain (lowers the pain threshold) and may serve the natural purpose of alerting the body to disturbances of its normal functions. Another prostaglandin promotes coagulation of the blood; another contracts the uterus, and so forth. All of the natural prostaglandins are made in body cells from a long-chain molecule called arachidonic acid by the action of successive enzymes. Like water falling over rocks, the intermediate substances and the final products cascade from the arachidonic starting material. Aspirin and the other nonsteroidal drugs act to block one of the enzymes (cyclooxygenase) in this cascade, which leads toward inflammation in the body. Because the cyclooxygenase step is blocked, the inflammatory prostaglandin cannot be made, and swelling, heat, and pain are prevented. This is a wonderful example of how research to explain the action of a natural biochemical system can lead to practical results in quite another area. Because it was learned that these drugs block cyclooxygenase, a simple screening method using this enzyme in test tubes avoids the necessity of using animals to find the prostaglandin—an improvement over the previous tests.

Another equally effective type of drug to relieve headaches and other moderate pains is represented by acetaminophen, which works in a different way from the acidic antiinflammatory medicines. Many patients who fear the effects of acidic preparations on the stomach prefer acetaminophen, which is the active ingredient of most of the "aspirin-free" nonprescription painkillers. The available brands are distinguished by pill size, enteric coating of tablets to prevent their solution in the stomach, and beautiful colors—which, of course, do not contribute anything to dull the pain. The only thing that such beauty treatments ensure is a steep increase in price.

Aspirin, acetaminophen, and the other nonsteroidal antiinflammatory analgesics relieve only minor types of pain. These drugs do not cure the causes of pain, and they give only temporary relief of symptoms, yet the volume of their sales is extraordinary. Figures of $1 billion per year may be low estimates.

Superficial or surface pain from neuralgias is believed to be helped by local application of oil of wintergreen and similar liniments. Counterirritants are agents applied locally to produce superficial inflammation that reduces inflammation in deeper adjacent structures; they can also make one forget such pain for a short time. The pain of itching and skin allergies can be eased by applying hydrocortisone and other steroids to the inflamed area.

The steroid drugs have lost some of their popularity because of their side effects. Cortisone and hydrocortisone (also known as cortisol) are hormones that were isolated from the *cortex* (outer layer) of the adrenal gland more than 50 years ago and have been synthesized commercially. Their antiinflammatory action cannot be separated satisfactorily from their many different and interdependent biological activities. At first nobody wanted to alter their chemical structure because of the prevalent belief that "nature knows best," so what chance would a mere chemist have to compete synthetically with chemicals selected by evolution over eons? However, as in the case of many other natural products, this idea became untenable when J. Fried and E. F. Sabo in 1954 made changes in some hormones and succeeded in preparing variants much stronger and somewhat more specific than cortisone and hydrocortisone. Since then, hundreds of such analogues have been tested, and as usual, a few of them survived the demands of clinical trials. To cite just three examples, prednisone, prednisolone, and dexamethasone are steroids widely prescribed for severe cases of rheumatoid arthritis and other inflammatory conditions.

Because the steroidal hormonal agents just mentioned occur in minute amounts in the cortex of the adrenal gland, they had to be synthesized on a large scale to become available as medications. In the commercial synthesis of cortisone, 38 separate reactions had to be performed, and the yield shrank at every step. Not surprisingly, the value of synthetic cortisone was many times that of the same weight of gold.

In the course of these operations, one team of chemists made a brew of a likely intermediate and poured it into a tank in which a common bacterial culture was maintained. The bacteria metabolized the added reagent; when the mixture was extracted and purified, the jubilant chemists found that the microbes had done exemplary chemical work, cutting out eight steps of the manual synthetic sequence.

Cortisone and its analogues are highly chiral compounds that can exist in several nonequivalent shapes. It took luck and chemical genius to isolate the correct enantiomer in good yield from the final reaction mixture.

ANTICHOLINERGICS

Belladonna drugs, which have been studied for more than 150 years, come from alkaloid-bearing plants. The name *Atropa belladonna* was coined by Linnaeus in the 18th century for the deadly nightshade vine to remind people of the poisonous nature of this plant known to writers of the Hindu Veda and to other authors in Roman and medieval times. Atropos was one of the Greek Fates who cut the thread of life. Large doses of extracts from belladonna were used to cause insidious and hard-to-trace deaths from its poison. Because weak extracts also dilate the pupil of the eye, women in the Renaissance used belladonna to make their eyes appear more brilliant, hence the name *bella donna* (which means beautiful lady).

The alkaloid atropine was isolated in 1831, but it was later recognized as a mixture of chemicals with mirror-image structures called optical antipodes (the hyoscyamines, named after the henbane shrub, *Hyoscyamus niger*). Atropine also occurs in *Datura stramonium*—called variously thornapple, stinkweed, and jimson weed (because it was found near Jamestown, Va.). Another important alkaloid, scopolamine, was named after the plant *Scopolia carniolica*, but it also occurs together with atropine in henbane.

Atropine is used widely in medicine. In small, nontoxic doses it slows secretion by glands and therefore helps in drying up unwanted fluids during operations. Dentists use it to control saliva. Atropine also acts as a powerful antispasmodic drug to prevent spastic contractions of involuntary muscles such as those of the throat, stomach, and intestines, over which people have little or no conscious control. This antispasmodic activity extends to muscles in the eye; therefore atropine has been used in the past to dilate the pupils to see inside the eye.

Scopolamine has similar properties, but these are overshadowed by its effects on the brain. In suitable concentration this drug depresses the central nervous system, causing twilight sleep, a state useful in obstetrics and minor surgical procedures. In smaller doses, injections of scopolamine used as a truth serum in interrogations produce the relaxation necessary to gain the subject's cooperation. Scopolamine is also marketed in skin patches for transdermal application. The drug traverses the skin and protects the bearer against motion sickness. Both atropine and scopolamine work by counteracting the neurohormone acetylcholine. Acetylcholine does many different things. Drugs with many different actions are not as valuable as those that do one or two things, and therefore thousands of attempts have been made to develop chemical relatives (synthetic analogues) of atropine and scopolamine that have fewer actions. Because these alkaloids have a reasonably simple ester type of structure, chemists have altered one section of their molecules after another in these attempts. However, none of the antispasmodics found have proved appreciably better than the natural ones. Somewhat more specific drugs have been developed as dilators of the eye (*mydriatics*), such as homatropine. The main advantage of some of these synthetic and less powerful anti-spasmodics is their decreased depression of the central nervous system. As always, drugs with specific actions and few side effects are ideal for clinical uses.

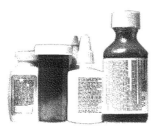

10

Local Anesthetics, Antispasmodics, and Antihistamines

LOCAL ANESTHETICS

Local anesthetics are used to block the transmission of pain messages to the brain without producing the unconscious state resulting from general anesthesia. These drugs perform well when injected into the gum in dental surgery, into muscles and under the skin for minor surgery, or dropped into the eye to block pain there. They can also be injected into various sections of the spinal cord, where they block the awareness of pain sensations coming from whole sections of the body. A few local anesthetics perform well through the skin and on the lining of the nose and mouth (the mucous membranes). These drugs counteract itching and other surface pain. Dentists use them to numb the surface of the gum to avoid pain from subsequent injections.

For ages, people have rubbed all kinds of leaves, oils, and other materials on themselves to blunt topical pain, but numbing underlying tissues was first achieved with cocaine. The dangers of cocaine prompted medicinal chemists to search for other local anesthetics. They thought it likely that the molecular structure of cocaine determined its anesthetic activity, and this idea guided the planning of their research. Piece by piece they chipped away sections from the cocaine molecule until they found structures with local anesthetic activity. Among these drugs were benzocaine and procaine. For a while, procaine became the most widely used injectable local anesthetic, best known by its proprietary name, Novocain. Benzocaine was simpler; it has survived for eight decades and is still used on surface tissues.

The success of these drugs, especially procaine, opened floodgates of imitation. Hundreds of similar local anesthetics were made, tested, used, and forgotten over the years. By silent consent, their names all ended in -caine. The French chemist Ernest Fourneau in 1904 translated his name into English (furnace or stove) and named his variety of local anesthetic in his own honor, stovaine. Most of these drugs had the same drawback: They did not last long enough in the tissues and had to be renewed by a drip technique. Their disappearance is caused by a part of the molecule called an ester group. Ester groups are decomposed by esterase enzymes. Therefore, the chemists explored other, less easily destroyed variants that had ether and amide groups. One of these variants proved to be a long-lasting surface anesthetic (dimethisoquin); others, lidocaine (Xylocaine) and mepivacaine (Carbocaine) are widely used, long-lasting, injectable local anesthetics.

Local anesthetics act by blocking microscopic channels through which mineral ions (such as sodium, potassium, and calcium) pass to the interior of nerve cells. The first step in the complicated sequence leading to this blocking appears to be a chemical interference with the nervous transmission of impulses at specific places (nodes of Ranvier) on parasympathetic nerves. Other drugs, such as antispasmodics, block pain messages through the parasympathetic nerves at other locations.

ANTISPASMODICS AND ANTIHISTAMINES

Antispasmodics block the neurohormones, acetylcholine and sometimes serotonin, from transmitting nervous impulses. They keep muscles from becoming spastic and glands from excessive secretion. Among other actions, they relax the muscles that contract the stomach and intestines. Chemically they resemble biogenic amines, but at one end of their molecules they have one or two bulky, inert structures that act as blocking groups. When, for example, acetylcholine is attracted to receptors, if an antispasmodic is present, the neurohormone is blocked physically from joining its receptor. On a molecular basis, this activity is similar to what happens when a football player running for a goal meets two big opponents who block his way and pull him down.

Many antispasmodics are available to physicians, led by the ancient alkaloid atropine, which is still one of the most powerful drugs of this type (*anticholinergic*). When given before an operation, it dries up glandular secretions and relaxes involuntary muscles.

In a similar type of interference, the action of histamine at its receptors (called H-1 or histamine-1 receptors) is blocked by antihistamines or, as they are now called, H-1 receptor antagonists. These drugs prevent histamine from linking to its receptors by blocking structures similar to those of the bulky antispasmodics.

It is easy to understand why drug activities overlap. For example, the antihistaminic drug diphenhydramine not only efficiently counteracts histamine-caused allergic reactions such as running nose and hay fever, but it also controls muscular spasms as well and even sedates the central nervous system and induces sleep. In fact, many over-the-counter drugs owe their sleep action to diphenhydramine. By slowing nervous transmission in the cochlear region of the inner ear, the same drug counteracts motion sickness. Travelers can take Dramamine, which is a different salt of diphenhydramine, to prevent sea or air sickness.

Approximately 30 antihistaminic drugs are available for medical treatment. Most of them cause drowsiness and should be avoided by people who must drive an automobile or run machinery. But not all people are sedated by the same agent. Some people show no sign of sedation when taking the same drug that puts others soundly to sleep. Physicians and patients must experiment to find out which drug a given patient tolerates best.

Histamine has another activity that does not respond at all to the typical antihistamine: it can increase the secretion of hydrochloric acid by the mast cells of the stomach. (Mast cells are large cells with numerous heparin-containing granules that occur especially in connective tissue.) The stomach lining protects itself from the harmful effects of hydrochloric acid and the enzymes contained in pepsin of the gastric juices by various defense responses, such as mucus and bicarbonate secretion and regeneration of the cells of the stomach and duodenal walls. If these protective measures fail, peptic or duodenal ulcers may develop.

Histamine activates the response to histamine-2 (H-2) receptors, which increase the secretion of hydrochloric acid. A number of drugs that were discovered in the late 1970s specifically block H-2 receptors and thereby the secretion of damaging components of gastric juice. Among these drugs are cimetidine, ranitidine, and famotidine.

Gastric ulceration has been traditionally regarded as a consequence of stress. Now an organism, *Helicobacter pylori*, has been cultured from the human stomach. It is suspected to cause gastric ulceration. It inhabits the epithelium of the stomach and the gastric mucus, and it renders the mucosa more vulnerable to injury. *H. pylori* has been found in 90% of patients with duodenal ulcer and 80% of gastric ulcer patients. It should be possible to reduce this bacterial infection with antibiotics. Indeed, triple therapy with tetracycline or amoxicillin, metronidazole, and bismuth salts such as Pepto-Bismol for two weeks eradicates 80–90% of the infection. Interest in *H. pylori* has increased because of the observation that infection with this organism may be involved in the occurrence of stomach cancer.

Research on additional types of histamine antagonists continues because histamine is suspected to be the "bad boy" among body chemicals. The release of histamine in brain tissues at so-called H-3 receptors may well cause ill-understood disturbances such as headaches, and drugs that block its action at central nerve cells might be of value in controlling such painful afflictions.

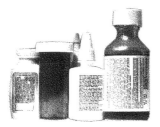

11

Drugs That Act on the Blood Pressure and the Heart

It is now possible to control both low and high blood pressure with drugs that bring pressure levels in line with more normal measurements. However, these medicines do not cure the causes of abnormal blood pressure. Low blood pressure (*hypotension*) occurs during periods of starvation, exposure to cold, loss of blood, and surgery. After a meal, blood flows to the tissue along the stomach, and this blood flow drains blood from the brain. In some individuals such a postprandial drop in the blood pressure may cause dizziness. If these difficulties are temporary, it may not be necessary to resort to drugs, but if they persist, a drug to increase blood pressure may be needed. In extreme cases of low blood pressure, epinephrine or its synthetic analogue, isoproterenol, which both constrict blood vessels, can be tried. Milder cases respond to amphetamine, ephedrine, and similar drugs.

The more serious and insidious change in blood pressure is high blood pressure, or *hypertension*. There are many causes of high blood pressure, and no one type of drug regulates it for everyone. Only a few of the types of hypertension that can now be treated will be discussed here. Some cases are caused by the constriction of blood vessels in the periphery of the body, others by the failure of kidney enzymes that regulate excretion of sodium. Still others are caused by changes in the regulating mechanisms in the central nervous system.

A threatening form of high blood pressure is called essential hypertension—a strange name that seems to say that everyone needs it, whereas the opposite is true. Not all of its causes are understood. Persistently high levels of blood pressure overwork the heart and endanger the blood vessels, including those of the brain, and contribute to heart attacks and to "cerebral accident," which is usually called stroke.

Epinephrine (Adrenalin) increases blood pressure by narrowing small blood vessels and building up pressure behind these constrictions. As long as this narrowing goes on in a normal way, all is well. The action of epinephrine must not be allowed to get out of hand, however. Epinephrine acts on two kinds of receptors, called alpha and beta adrenergic receptors.

Each of these types of receptors has been subdivided (into α_1, α_2, β_1, β_2, etc.) on the basis of their location in different parts of the body (such as the heart, lung, and brain). Joining epinephrine with the alpha receptors increases the force with which the heart contracts and thereby increases the blood pressure. Isoproterenol reacts at beta receptors and can revive a failing heart. Irregular heartbeats, called *cardiac fibrillation*, can be brought back to a normal rhythm by blocking the beta receptors.

The development of beta blockers (technically called adrenergic beta receptor antagonists) was planned and accomplished by Sir J. W. Black in Great Britain in 1962. More than 400 analogues were tried, and finally propranolol was chosen to control irregularities of the heartbeat. Later, scientists learned that propranolol also lowered high blood pressure effectively, acting on peripheral nerves and in the brain. Still later, it was observed that this drug reduces pressure in the eye from glaucoma.

Propranolol's interesting health benefits and high sales figures set competitive research in motion in many large pharmaceutical companies. Two of the resulting drugs are nadolol, which is not broken down to any extent in the body but is excreted slowly and therefore acts for a long time, and timolol, which is specific for the treatment of glaucoma. Nadolol can also prevent migraine headaches.

A similar late recognition of pressure reduction occurred in the history of thiazide diuretics, which were originally designed to treat edema by improving the excretion of water and salt by the kidney.

The thiazides are often used in combination with other pressure reducers such as beta blockers.

Epinephrine is the acknowledged cause of some forms of high blood pressure, and it was only natural for chemists to try to block its synthesis in the body. One drug resulting from these efforts is methyl-dopa, which lowers the blood pressure in some cases by interfering with some of the enzyme catalysts needed to make epinephrine.

Every nerve path involved in changes in blood pressure has been studied, and drugs designed to act at these locations have been tried. Nerve cells cluster around nerve ganglia, which serve as a sort of telephone exchange. Ganglionic blocking agents have been made, tested, and often abandoned. Reserpine still has restricted uses in treating high blood pressure. One reasonable suggestion—to relieve the blood pressure built up behind narrow blood vessels with drugs to dilate the vessels—has had only limited success.

This short selection of medicines to control high blood pressure is intended to give the reader an idea of the variety of approaches, the groping for rational treatments, and a few selected accomplishments in this difficult and incomplete area of medicinal research.

DRUGS FOR HEART DISEASES

The heart muscle contracts and relaxes rhythmically under the influence of a number of biochemical reactions. Its rhythm is controlled by involuntary (*autonomic*) nervous impulses. The reactions that power the contractions are mostly exchanges of ions, that is, electrically charged fragments of molecules and atoms. Among them are sodium, potassium, and calcium ions, which flow across cell membranes or, at times, have to be made to flow across them. The chemicals that do the transporting are adenosine triphosphate (ATP) and its chemical relatives. Their reactions release energy both as heat and as electricity. This energy is used to convey ions across membranes. Once across, the ions activate proteins in the heart muscle, which then gather together into bundles of fibers. These bundles, in turn, contract and then relax, returning to a resting position. This process is much more complicated than the outline given in these few words. Drugs that stimulate the force of the contractions are called positive inotropic agents (*ino* means muscle, and *tropic* means to influence in Greek).

The classical inotropic drugs are the glycosides of digitalis. Dried and powdered, the purple foxglove plant, used since the 13th century, was often toxic. In 1785, the British physiologist William Withering showed the inotropic value of digitalis and, by standardizing plant samples and adjusting doses to each patient's needs, recommended ways to reduce its danger. Standardization proved too much trouble for most physicians, and it took another 150 years before digitalis became the accepted medication for hearts that did not contract forcefully enough.

Most physicians prefer to prescribe a purified, well-standardized drug that is now obtained from the woolly foxglove (*Digitalis lanata*) and is called digoxin (Lanoxin). After patients have taken higher doses to establish adequate levels in their blood, they can be maintained on low daily doses.

The contraction of the four heart chambers must not only occur forcefully but also with great regularity in a rhythmic fashion. This process requires the coordination of a complicated system of electric discharges, like a pacemaker that regulates the automatic cooperation between the chambers. The two atria at the top of the heart must contract and empty their contents before contraction of the ventricles at the bottom begins. If any of the four fail, the rhythm is broken. Atrial irregularities (*arrhythmias*) are not regarded with the same dread as ventricular fluttering (*fibrillation*), which occurs when secretions of the lower muscles die after the arteries that supply them with blood have become clogged. Drugs that prevent irregular contractions are called *antiarrhythmics*. More than half a dozen of these drugs are in use today. Among them is quinidine, an alkaloid occurring in cinchona bark together with its enantiomer, the antimalarial drug quinine. Unfortunately, quinidine is not well tolerated by many heart patients. Like other nonspecific antiarrhythmic drugs, quinidine stabilizes the membranes of heart cells by checking the flow of certain ions through them, thus reducing their permeability.

Other nonspecific drugs apparently work the same way and help some patients by stabilizing irregular beats. Two of these drugs are local anesthetics, lidocaine and procainamide, which are used widely during heart attacks. Phenytoin, designed to control convulsions, also stops fibrillation. In all these cases, the action on the heart was discovered after the drug had been introduced for other uses.

Propranolol and a number of related drugs were designed to control heart rhythms, tested for that purpose, and introduced into cardiology as useful tools. Their usefulness in other ways, such as lowering high blood pressure and preventing painful attacks of angina pectoris, was discovered later. On the other hand, the drug verapamil was tested for relief of angina during its development and studied with special attention to its action on the flow of calcium ions through the calcium ion channels in the membranes of cells of the heart.

Calcium ion blockers block transport of calcium through the cell membranes. Their action is an example of medical interference with the action of enzymes. One such drug is digitalis, which inhibits the enzyme sodium potassium ATPase. Sodium potassium ATPase transports calcium ions into the cell and sodium ions out of the cell into the surrounding serum. This reaction in turn leads to an increase in free calcium ions, which strengthens the contraction of the heart muscle, thus easing congestive heart failure.

Calcium ion channels control the entry of calcium ions into the cells of the heart muscle from the space outside the cells. Inhibitors of the transport of calcium ions through these microscopically narrow channels, such as verapamil, nifedipine, and diltiazem, decrease the contractibility of the cardiac muscle and are therefore useful in the treatment of hypertension and of angina pectoris.

The first effective drug to relieve anginal pain was the liquid amyl nitrite, which has been available since 1867. Some more stable nitrogen compounds have replaced amyl nitrite, especially nitroglycerin (glyceryl trinitrate). This high explosive, the active component of dynamite, opened the era of modern blasting and earned the Swedish mining engineer Alfred Nobel the fortune that still finances the Nobel Prizes.

Nitroglycerin used to be sold in capsules that were placed under the tongue and dissolved slowly. It is now available in small sublingual tablets (Nitrostat) that act more quickly. The nitrate esters of mannitol, isosorbide, and some other carbohydrates can be swallowed and act after reaching the digestive organs, but that takes more time.

The way drugs regulate the force and rhythm of the heartbeat is bewildering to the general reader. The acquired experience of a cardiologist is of help in deciding what therapeutic procedure should benefit a given condition. The automatic heartbeat may fail if certain enzyme systems, each consisting of several cooperating enzymes that keep the heart on its regular schedule, deteriorate. The sequence of events controlled by the regulatory enzymes and the ion channel transporter compounds is not yet fully understood. The toll that heart diseases and the associated diseases of blood vessels (*thromboses*) throughout the body exert on mortality rates and on the quality of life of the survivors has made research on cardiovascular drugs a priority through the world.

DIURETICS

Diuretics not only increase urine flow but also help the excretion of excess salt and ions by triggering their selective filtration in the kidney. The fluids of the mammalian body are similar to seawater, from which all animals arose. Their concentration of salts, especially ordinary table salt (sodium chloride), is one-fourth that of seawater, namely, about 0.23%. This percentage is the same for blood, lymph, tears, saliva, and stomach fluid. Solutions with equal salt content are called *isotonic*. Some body fluids are acidic (such as stomach juice); some are alkaline (such as intestinal juice and saliva); others, like blood serum, are neutral.

Blood circulates through all body organs, including the kidney. The kidney contains yards of fine tubes coiled inside. In the kidney, the blood is separated into its components by a complicated process of absorption and filtration. Each kidney cell (*nephron*) lets water and electrically charged atomic particles (*electrolytes*) pass into the tubules while returning blood corpuscles and nutrient molecules to the circulation. Not all charged particles are allowed to pass through the kidney tubules. This selective filtration keeps enough electrolytes in the blood to maintain its normal isotonic concentration. If the blood contains too much blood sugar for extended periods, the excess glucose passes through the kidney and appears in the urine. Proteins do not pass through kidney membranes unless the kidney is infected or otherwise damaged. The appearance of protein in a cloudy urine is a sign of kidney disease.

Salts must dissolve in about 99% water to stay at isotonic concentration. If excess ions—of sodium, for instance—are to be filtered out and discarded in the urine, corresponding amounts of water must accompany them. Likewise, if excess ions accumulate in the body, the body's tissues swell because of the water that holds the salts. This condition is called edema and occurs during kidney diseases, toxemia of pregnancy, premenstrual tension, and as a side effect when steroids are taken for a long time. The most common cause of edema is congestive heart failure, in which the heart cannot pump blood through the kidney well enough to allow the kidney to filter properly.

Stirring the kidney to greater activity (*diuresis*) can be done by many chemicals. Several of them are only historically interesting. Drugs containing mercury, such as mercaptomerin and meralluride, are still used occasionally. Caffeine, which is found in coffee, tea, and cola drinks is mildly diuretic but is not used clinically. It belongs to a class of chemicals called xanthines, of which theophylline is more effective as a diuretic. Theophylline is usually combined with other, more active substances.

The drug triamterene can allow sodium to be filtered out into the urine while leaving potassium behind. Usually these two ions travel together. In combination with the more powerful sodium-excreting drugs, triamterene has filled an important place among kidney drugs.

Probably the most widely used diuretics stem from the thiazides; chlorothiazide and hydrochlorothiazide were the first of these drugs to be prescribed. In addition to being diuretics, they lower the blood pressure, a valuable aid in congestive heart failure. The thiazide studies followed the discovery that certain sulfa drugs (sulfanilamides) block the enzyme carbonic anhydrase, which, in the kidney, changes the acidity of the urine. Blocking this reaction increases the flow of urine. The first successful diuretic that blocked this enzyme was acetazolamide, which was synthesized and tested by Richard O. Roblin, Jr. The drug also blocks the same enzyme in other organs, such as in the eye, where it reduces the pressure from glaucoma. The thiazides are clever chemical relatives of the carbonic anhydrase blockers, but they act by a variety of different mechanisms. They were first developed by a team of industrial scientists, James M. Sprague, F. C. Novello, and K. H. Beyer at Merck, Sharp and Dohme.

Thousands of sulfa compounds were tried, and furosemide was found in the Hoechst Laboratories in Germany. It is a *high-ceiling* diuretic, that is, it acts by inhibiting sodium and chloride transport in the ascending loop of Henle in the kidney. Ethacrynic acid increases the flow of urine by acting on kidney filtration as a loop diuretic.

Blocking any hormone that tends to retain fluid in the body would seem to be a logical idea to increase the flow of urine, but this is easier said than done. In the case of aldosterone (an antidiuretic hormone made in the adrenal cortex), these efforts produced the drug spironolactone, which bars aldosterone from its receptors.

ANGIOTENSIN-CONVERTING ENZYME INHIBITORS

Another way to lower pathologically high blood pressure is based on the role of several enzymes. One of them, the kidney enzyme renin, acts on a peptide called angiotensin I that circulates in the blood. Angiotensin I is rapidly converted to another peptide, angiotensin II, by an enzyme-catalyzed process that gets out of hand in cases of disease. Angiotensin II is the most potent vasoconstrictor known; it drives up the blood pressure if it is not stopped. Thus, prevention of the release of renin, which starts the whole scheme that produces the hypertensive angiotensin II, or inhibition of the conversion of angiotensin I to angiotensin II could prevent the feared rise in blood pressure. A team of medicinal chemists and pharmacologists at the Squibb Institute for Chemical Research led by M. A. Ondetti and D. Cushman developed the drug captopril, which inhibits the enzyme that converts angiotensin I to angiotensin II; captopril thereby effectively lowers the blood pressure.

Altogether, research on diuretics and specific and pertinent enzyme inhibitors has given physicians a variety of drugs that relieve hypertension, edema, kidney failure, and congestive heart failure.

COAGULANTS AND ANTICOAGULANTS

Blood circulates through large, small, and minutely small vessels. Its fluidity is maintained by numerous biocatalysts that keep blood from clotting within the vessels and that keep it from leaking through the vessel walls. Unwanted clotting may lead to thrombosis, and improper leakage results in hemorrhage. Coagulants prevent hemorrhage; anticoagulants prevent thrombosis.

Blood clotting takes place in several phases, in which prothrombin is formed from plasma proteins that are labeled by Roman numerals, for example, factor VIII. Prothrombin is converted to a short-lived catalyst called thrombin. Under the influence of thrombin, the water-soluble protein fibrinogen is polymerized and forms an insoluble fibrous material called fibrin. Blood corpuscles become enmeshed in fibrin, and this mixture is the blood clot.

This highly simplified representation of the clotting process might suggest that it can be speeded up or interrupted at many stages, and indeed, this is true. Vitamin K—which is present in such foods as alfalfa, spinach, cabbage, soybean oil, and egg yolk, and is also synthesized by friendly intestinal bacteria—aids some of the early steps of blood clotting. Conversely, persons who lack factor VIII suffer from hemophilia, a serious, inheritable bleeding disorder.

There are several K vitamins; one of them with a simple chemical structure is called menadione. It is used to stop bleeding in surgery and other traumatic events. There are water- and oil-soluble derivatives of menadione. Vitamin K does not enter into the proteins of blood coagulation but acts as an essential cofactor.

For minor cuts that cause bleeding, astringents such as alum can be used. More serious bleeding may require a corticosteroid, aminocaproic acid, or other prescription drugs.

Two types of anticoagulants are in clinical use. One is injectable heparin and related products, the other is dicumarol and similar substances that can be taken orally. These anticoagulants are used to prolong blood-clotting time in patients whose blood vessels have already been clogged by a clot after a cardiac or cerebral infarct (an area of dead tissue resulting from obstruction of the local circulation by a floating or stationary clot).

Heparin is an unusually structured polysaccharide not unlike starch but solubilized by water-related groups. It was discovered in 1916 by J. McLean, who was then a medical student at Johns Hopkins University. It is extracted from beef lung or liver; it has to be injected and cannot be administered otherwise. It inhibits the formation of thrombin from prothrombin and thereby prevents the clotting mechanism.

The need to inject heparin makes it inconvenient to use for patients who have to take an anticoagulant continuously. It was therefore a fortunate circumstance when the orally active anticoagulants of the coumarin type were developed. The path of this discovery is interesting. In 1922, F. W. Schofield called attention to a serious disease of cattle in the upper Midwest United States. The animals suffered from internal bleeding that was traced to their fodder, improperly cured sweet clover hay. K. P. Link at the University of Wisconsin isolated the responsible toxic product and later synthesized it. He called it dicumarol. The compound was introduced as a clinical anticoagulant in 1941.

Molecular modification of this substance provided several more or less related anticoagulants. Only two will be mentioned. One is diphenadione, a drug that acts like dicumarol by preventing the bio-synthesis of prothrombin. Diphenadione must be given with caution because hypersensitivity and bleeding may occur. The other oral anti-coagulant is warfarin. It is also used as a rat poison. This compound was tested routinely in laboratory rodents when the high mortality of the animals due to internal hemorrhage was discovered. This story suggests that its use as a rodenticide in the warfare against rat damage to grains in commercial silos gave it its name.

12

Intestinal Tract Medications

Several classes of drugs affect the stomach and intestinal tract, but only three types will be discussed in this chapter: antacids, cathartics, and constipating drugs.

ANTACIDS

Stomach antacids lower the acidity of the stomach (*gastric*) contents. Too much acid (*hyperacidity*) is evident in heartburn or hiatal hernia, in which a break in the diaphragm, which divides the upper chest from the stomach, allows some fluids from the stomach to pass back up instead of all moving downward. Acidity is also apparent in inflammation of the stomach or pancreas, in gallstone problems, and in cases of angina pectoris. The excess stomach acid (hydrogen chloride) can be neutralized with baking soda (sodium bicarbonate) or with calcium carbonate, which goes into solution more slowly. Milk of magnesia (8% magnesium hydroxide) and other magnesium compounds form the active ingredients of the many tablets that executives take on television when the company's profits are down. The treatment of hyperacidity with histamine-2 receptor antagonists was described in Chapter 10.

CATHARTICS

One of the oldest cathartics is castor oil, which is obtained by cold-pressing the seeds of the Euphorbiaceae, *Ricinus communis* L. Castor oil has been abandoned in most modern countries as a medical laxative but is widely used as an industrial raw material for coatings, hydraulic fluids, rubber preservatives, and as an embalming fluid.

One of the causes of constipation (*intestinal atony*) is loss of the successive, involuntary movements of the intestinal muscles, called *peristalsis*. Stress, faulty diet, dehydration, and prolonged use of constipating drugs may lead to peristaltic slowdown. Many people suffering from constipation reach for over-the-counter laxatives (from the Latin *laxare*, meaning to loosen) or cathartics (from the Greek *kathartikein*, meaning to purify). These drugs irritate the intestinal wall and thereby increase its movements. Some of them are inorganic salts such as sodium sulfate; Epsom salt (magnesium sulfate), the favorite cathartic of our grandparents; and milk of magnesia. Others are organic compounds; some of these compounds were found originally in herbs, and others are purely synthetic.

A widely advertised laxative is phenolphthalein. It has been known as an acid–base indicator since 1880. In acids it is colorless, and in bases it is red; this trait is useful for adjusting solutions to neutrality. The chemist Z. von Vámossy used phenolphthalein as an indicator of the acidity of cheap wines, as required by the Hungarian government in 1990, and found that it was not toxic and had no effect on dogs. In humans, however, it caused soft stools. Extended studies led to its introduction as a relatively harmless cathartic. It is tasteless itself and is sometimes coated with chocolate; it damages no vital organs; and it activates the muscles that move the large intestine.

Among botanical cathartics that are widely used in Europe and Asia are several anthraglycosides that occur in leaves, roots, and pods of aloe, frangula, rhubarb, senna, and other plants. The properties of some of them or their decoctions were known to ancient Arabs, Africans, West Indian witch doctors, and California Indians. The senna products are produced commercially; they are mild laxatives. Aloe species in Curaçao and eastern Africa and South Africa furnish similar drugs. Many people prefer bisacodyl, a synthetic agent whose chemistry is reminiscent of that of antispasmodics or other anticholinergics. It is a reliable cathartic.

CONSTIPATING DRUGS

The constipating chemicals have the opposite action from that of the cathartics. Loose stools and diarrhea may be caused by infections, by other diseases of the intestines, by overeating, or by cancers, or they may have psychosomatic causes. Overactivity of the colon and other intestinal segments may be toned down by loperamide or by diphenoxylate, drugs developed in the 1950s by Paul A. J. Janssen in Belgium. The constipating effects of the complex activity of opium, paregoric, morphine, and codeine are well known. Materials that absorb other chemicals, fill the colon, and make intestinal contents thicker are used by many people because of their lack of side effects. Kaolin, pectin (both of which are in Kaopectate), and various bismuth salts (as in Pepto-Bismol) are among these preparations. None of them removes the causes of the intestinal overactivity, but they offer temporary relief from annoying symptoms and malaise. Infections leading to long-term diarrhea, however, must be treated with antibiotics.

13

Hormones and Vitamins

SEX HORMONES

Extremely small amounts of sex hormones were isolated from the testes, from urine of pregnant animals, and from the corpus luteum ("yellow body," an endocrine gland that secretes progesterone) and ovaries around 1930. The chemistry of these compounds was studied and brought Nobel Prizes to a string of organic chemists. All of these hormones are steroid derivatives produced from cholesterol in the sex glands and auxiliary glands. The principal female hormone, estradiol, is broken down (*metabolized*) by oxidation, and its oxidation products, such as estrone, are excreted in the urine. Chemists processed 40,000 liters of urine from pregnant mares to isolate a few milligrams of estrone. Estradiol and estrone make female animals go into heat (*estrus*). In women, the estrogens induce the menstrual cycle and ovulation.

The other type of female hormone is progesterone. Its source is the corpus luteum; its mission is to prepare the uterus to receive and embed the fertilized egg (*ovum*) and to maintain pregnancy to term.

The male sex hormone, testosterone, is needed for the maturation of the male sex organs, for the production of sperm, and for the maturation of secondary male characteristics. Of these characteristics, change of voice and growth of beard and body hair are called *androgenic* effects, while development of male muscles is called the *anabolic* effect. Muscle development can be increased and separated from sex hormonal activity by some synthetic derivatives of testosterone. These anabolic steroids are taken by mouth by some athletes to build up muscle mass, but this usage involves health hazards, such as masculinization in women, edema, fever, jaundice, local irritation, and some forms of impotence.

Testosterone is metabolized by the body in various ways; one of them is reduction, that is, addition of hydrogen atoms. The body achieves this process with the aid of an enzyme, 5α-testosterone reductase. The resulting dihydrotestosterone has biological properties similar to those of its parent hormone, but it also enlarges the prostate gland, which straddles the urethra between the bladder and the penis. Most older men suffer from an enlargement of the prostate gland and develop delays and other unwanted symptoms of urinary flow.

By inhibiting the reducing enzyme, 5α-testosterone reductase, the enlargement of the prostate gland can be slowed down and even reversed in some cases. Several such inhibiting drugs have been developed, for example, finasteride (Proscar). The drug terazosin relaxes the tight muscles of an enlarged prostate gland without affecting the continued growth of the gland. Its use, though, is beset by side effects of pronounced low blood pressure.

Some cancerous growths depend on hormones. The growth of breast cancer is enhanced by estrogens, that of prostate cancer by androgens. In reverse, cancer of the breast is retarded by androgens, and prostatic cancer by estrogens. Molecular modification of the natural sex hormones has produced structural variants that are preferred for treating specific problems. Testosterone and some of its analogues have been implicated as causes of acne and male baldness.

Totally synthetic estrogens have also been made. One of these drugs, diethylstilbestrol, known as DES, was used by millions of women as a fertility hormone. It was also used in the 1950s to prevent miscarriage. The children of mothers who used DES are now showing abnormally high rates of uterine and testicular cancer. Tons of this compound have been used to enhance the growth of farm animals, but with unexpected results. To be sure, it is cheaper than estradiol, but it remains in the food chain and can initiate malignant cancers in people who eat the meat of treated animals. An analogue of this drug, called tamoxifen, is used in the treatment of breast cancer in premenopausal women. It may block estrogen receptor sites.

The greatest success of molecular modification in this area has come from synthesizing analogues of progesterone. The natural hormone must be injected because progesterone is not active when taken by mouth. A synthetic variant with two slight changes in the original structure is active as a pill. It is used by women to imitate the conditions of pregnancy, in which fertilization cannot take place. Thus, a sexually active woman is protected from becoming pregnant. This is the principle underlying the contraceptive pill. To approximate the conditions of monthly hormonal changes and keep a woman on her usual nonpregnant chemical course, a weak estrogen (such as estrone methyl ether) is incorporated in "the Pill."

Prevention of pregnancy also lowers the need for abortion. Several antiprogestins (progesterone antagonists) act by denying the fertilized ovum its attachment to the uterine wall. The best known of these compounds is mifepristone, known as the "abortion pill." It has been studied in the Roussel Uclaf Laboratories in France and has become known under its experimental number, RU 486.

The commercial manufacture of contraceptive steroids could not have been accomplished if some abundant steroids from Mexican yams had not been found to serve as starting materials for relatively short, partially synthetic procedures. This finding also lowered the price of contraceptives to generally acceptable levels.

INSULIN

Insulin is a hormone secreted by the beta cells of a tissue in the pancreas. Its function is to regulate the level of blood sugar (*glucose*). Lowering the normal amount of glucose leads to *hypoglycemia*. Because all organs, especially the brain, depend on glucose for energy, too little blood sugar interferes with normal, energy-requiring activities. When glucose levels are raised—as after a meal—high blood sugar (*hyperglycemia*) results. In normal individuals, glucose returns to average concentrations (100 mg in 100 mL of blood) quite rapidly, but if something goes wrong with the supply of insulin, glucose levels remain high and the extra glucose is excreted in the urine. This excretion happens in diabetes.

Insulin keeps the level of blood sugar normal in several ways. One way is its effect on the passage of glucose molecules through the membranes that enclose tissues. Ordinarily, glucose can cross tissue membranes easily. It is stored in the tissue cells after being polymerized into a starchlike material called *glycogen*. This storage process removes glucose from the circulation. Part of the circulating glucose is also oxidized or burned with the catalytic help of insulin, which lowers the levels of blood sugar. If there is not enough insulin, too much glucose remains in the blood.

Insulin is a medium-sized protein with an amino acid composition that differs slightly in different animals. In spite of their molecular complexity, artificial insulins have been synthesized in England, China, and the United States. Many related proteins have also been made in the laboratory. One of the practical aims of these experiments has been to prepare a hormonelike protein that can be taken by mouth instead of having to be injected, but so far this goal has not been reached.

Diabetic patients, especially those who have suffered from diabetes since childhood, must replenish their inadequate insulin by injecting commercial insulin daily or by carrying a surgical implant of a device that releases insulin in a regulated fashion. Insulin cannot be taken from human glands; therefore, animal pancreas glands (from cows, pigs, and sheep) are extracted instead. Unfortunately, there are occasional allergic reactions to animal insulins, but human insulin has not come on the market until recently. This new supply is a triumph of genetic engineering by scientists applying the recombinant DNA technique. Specially treated yeast or bacterial cells (*Escherichia coli*) are exposed to human pancreatic beta cells and adopt their genes (portions of nucleic acids) that make insulin. These genes, when incorporated into the bacterial genetic machinery, make the bacteria behave, in part, like the human cells. From then on, they make large quantities of human insulin, which can be purified and used to replenish the insufficient supply in patients.

In an unrelated series of events, French chemists noticed that treating animal infections with some of the sulfa drugs (sulfanilamides) lowered the levels of their blood sugar. By systematically changing the structure of these antibacterial drugs, German chemists developed compounds that were no longer antibacterial but had become antidiabetic. These sulfonylurea drugs can be taken by mouth; they improve the kind of diabetes that begins at an advanced age but are ineffective against juvenile diabetes. Like so many other drugs, the sulfonylureas have some undesirable side effects, but their clinical advantages generally outweigh these drawbacks.

OTHER HORMONES

The hormones of the thyroid gland are relatively simple amino acids containing iodine. The two most active of these hormones are thyroxine and triiodothyronine. They are made by the thyroid gland from iodide salts in the diet. People who live where iodine in soil and water is scarce do not get enough iodine in their food and suffer from goiter and other deficiency diseases, which today are largely prevented by the addition of iodide to table salt. Medical knowledge of thyroid hormones has advanced to the point where their production, action, and receptors are understood. Among other activities, these hormones help to control the rate at which the body uses energy. People with too few thyroid hormones are sluggish, while those with too many are overactive.

Hormones of the pituitary gland and other brain components are small proteinlike substances called *peptides*. They are composed of amino acids arranged either in lines or in rings. Small differences in these arrangements make a lot of difference in their activities. These structural variations suggest that the hormones evolved in different ways to meet the individual needs of various kinds of animals.

Several of these small hormones have been synthesized. By incorporating radioactive isotope atoms into their amino acids, such as carbon-14 instead of the usual carbon-12, it is possible to follow the path of a hormone through the body. The blood carries it to the tissues containing its receptors. If these tissues are laid over an X-ray film, the radioactive hormones leave black spots on the film, thus pinpointing the exact location of their action. The pituitary hormones contract the uterus, change the blood pressure, regulate calcium metabolism, and stimulate certain glands to secrete other hormones.

The growth hormone, also a peptide, can now be used to help the long bones of unusually short children to grow more normally. Growth hormones were exceedingly difficult to obtain until a few years ago because they came from cadavers only. They have now become available by recombinant DNA techniques. In the United States, 7000–8000 children now have their growth-hormone deficiency (*dwarfism*) corrected.

The hormones of the adrenal gland have been mentioned before. Lately, nerve cells and endings have been recognized as sources of neurohormones, not unlike glandular tissues in that respect. The pineal gland, which regulates circadian (daily) rhythms in the body, contains a hormone called melatonin; this compound is a close relative of serotonin. It has been found to induce sleep therapeutically and to counteract jet lag.

VITAMINS

Vitamins are compounds that are needed, usually in small quantities, for numerous metabolic processes in the animal body. If not enough of a vitamin is present in the diet, serious deficiency diseases may result. Like the hormones, vitamins are catalysts, but there is a difference; hormones are made by the body, whereas vitamins come from outside, either from the diet or, in some cases, from beneficial microbes that inhabit the body.

Some animals can make certain substances that are vitamins for humans. In such animals, these vitamins behave like hormones. For example, guinea pigs and humans cannot produce vitamin C (ascorbic acid), but rabbits can. Vitamin C is therefore a hormone in the rabbit.

Most of the vitamins are obtained from a normal mixed diet that includes meat, cereals, vegetables, fruit, dairy products, and inorganic salts. Vegetarians should make sure that their diets contain all the necessary ingredients for healthful living and growth. An ordinary diet in this country furnishes adequate amounts of vitamins, but in certain cases, when food is not assimilated normally or during stressful periods such as pregnancy, vitamin supplements may be required.

The vegetable kingdom is the principle source of vitamins in our nutrition. The K vitamins come from some vegetable sources but also from intestinal bacteria that synthesize them, and their hosts can then absorb them through the intestinal wall. Vitamin B_{12} comes from liver, and Vitamin E from wheat germ oil.

Almost all of the vitamins have been synthesized, and many are produced commercially. Synthetic vitamins are identical to the natural vitamins, but they are separate chemicals and not mixtures of various compounds, as may be encountered in natural sources. For example, synthetic vitamin E is alpha-tocopherol, whereas wheat germ oil contains a mixture of tocopherols, not all of which have vitamin activity. The B vitamins often occur together in foods, whereas individual B vitamins are usually synthetic and have to be mixed for complete vitamin B therapy. Common vitamin pills contain the required daily minimum dose of each vitamin. For special deficiencies, some more easily metabolized forms of certain vitamins are offered in extra-strength capsules. For instance, the consumption of folic acid in adequate amounts in the early days of pregnancy can decrease the occurrence of spina bifida, a neural tube defect, in the newborn.

Two B vitamins (riboflavin, or B_2, and cobalamin, or B_{12}) are synthesized by molds of the *Streptomyces* genus. The molds are supplied with nutrients that are precursors of the vitamins, and the molds ferment these nutrients to the desired end product. This fermenting process has greatly reduced the cost of these vitamins. Today, tons of vitamins are used to enrich foods. Vitamin C is manufactured by a multistep synthesis in automated chemical plants with computerized controls and robots supervised by a minimum of human operators.

The vitamins were named by letters in the sequence of their discovery from 1910 to 1950. As their chemistry unfolded, more explanatory generic names were given, such as thiamine for B_1, riboflavin for B_2, and pyridoxal for B_6.

The understanding of vitamins and their uses as medicines and for food enrichment have all but eliminated in this country such diseases as scurvy from lack of ascorbic acid, polyneuritis (beriberi) from absence of thiamine, pellagra from too little niacinamide and riboflavin, pernicious anemia from reduced ability to absorb cobalamin due to the absence of an enzyme called internal factor, rickets from failure to assimilate and use calcium and phosphorus normally because of a shortage of sunlight or the D vitamins, and xerophthalmia from want of retinol, or vitamin A.

Many other, less drastic deficiencies are helped by vitamin therapy. Considerable debate has surrounded claims that vitamin C prevents or cures colds and allergies. If these claims had not been supported by two eminent scientists, Linus Pauling (1901–1994) and Albert Szent-Györgyi (1893–1986), they would have been derided by the medical profession.

Ascorbic acid is a reducing agent; when taken in huge doses, hundreds of times the established amount needed for ordinary living, it reduces many oxidized compounds, including some that are known to cause cancer, and thereby should lessen their damaging effects. Although the evidence for this explanation is circumstantial, benefits from large doses of the vitamin cannot be denied. Excess ascorbic acid is excreted unchanged, and it appears to be nontoxic. This is not the case for large doses of the A and D vitamins, however. They disturb the metabolism of minerals. Some claim that vitamin E and a precursor of vitamin A, beta-carotene, retard the development of malignant tumors, but large amounts of these substances may be dangerous, especially for children whose overzealous parents give them excessive doses of these vitamins, often in the form of percomorphi liver oil. Symptoms may range from headaches to liver enlargement, bone deformation, skin eruptions, and others. Most symptoms disappear gradually on withdrawal of the vitamin.

How do vitamins and hormones produce their spectacular therapeutic effects in such small doses? One answer to this question is to be found in the participation of these biocatalysts in certain enzyme systems; for example, the B vitamin pantothenic acid is incorporated in a larger complex called coenzyme A, and this substance attaches itself to some enzyme proteins that need it for best catalytic activity.

Vitamins such as vitamin A and hormones such as the thyroid hormones also need specific receptors to which they can be linked. The receptors for these catalysts include proteins such as retinoid X receptors (RXR alpha and beta), which make it possible for DNA to recognize these hormones and vitamins efficiently. This specificity shows again that the action of chemicals as medicines is extremely complex and involves many stages and many chemical reactions before a tangible biological happening can occur.

14

Drugs for the Treatment of Cancer

Fifty years ago there was not one drug to suppress or cure any cancer; 30 years ago there were two, nitrogen mustard and methotrexate. Today, several dozen medicines are available to treat leukemias, some rare cancers, and even some widely found solid tumors. In a few cases, cures have resulted, especially in the rare and virulent choriocarcinoma and in Burkitt's lymphoma. The other chemotherapeutic drugs only slow down the multiplication of malignant cells. If such agents reach circulating cancer cells before the cells settle down in a tissue, they can retard growth so much that parts of the body's immune mechanism, the scavenger blood cells (*phagocytes*), can take up and digest the malignant cells and thus cure the incipient disease. This treatment works best when the disease is in its earliest stages. Unfortunately, diagnosis of cancer is seldom made at these stages, and chemotherapy becomes progressively less effective as cancers invade tissues and become better established. To be sure, the situation is not hopeless even in somewhat later stages. Surgery and radiation are used to remove or destroy cancerous tissue, but at a price; normal neighboring tissue is always affected. Combination of all three treatments, surgery, radiation, and chemotherapy, offers the best current hope of arresting the spread of cancers. It is against this background that anticancer drugs must be measured.

Cancer is not one disease but an estimated group of more than 100 diseases of different organs. For such a variety of diseases, there must be a variety of causes: environmental, dietary, and inherited dispositions to faulty metabolism. All these factors lead to one complete start, a permanent change (*mutation*) in a cell that causes errors in the directions for constructing genes or for making proteins. Genes that can trigger the change from normal to malignant cells have been identified and named *oncogenes*. Their composition and location on a chromosome has also been established. This long chemical stride forward in pinpointing a cause of cancers will undoubtedly speed the search for ways to stop the oncogenes from beginning their deadly work.

The ultimate causes of carcinogenesis remain uncertain. One reasonable suggestion is that the formation of free radicals—that is, unstable parts of molecules—causes a greater reactivity of various body chemicals, especially with oxygen. In this way, peroxides are formed, which aggressively attack otherwise sluggish and stable biochemicals. By altering these biochemicals, the peroxides initiate mutations. Evidence of this initiation is seen in the ability of antioxidants to decrease the chances for oxidative reactions. Vitamins C and E, which are established antioxidants, belong to this type of protective chemicals.

Some enzymes found in many body tissues work together to ward off free radicals. The enzyme superoxide dismutase, for instance, disarms the oxyradical by converting it to hydrogen peroxide. But because excess hydrogen peroxide can produce more free radicals, another enzyme, catalase, splits hydrogen peroxide into water and oxygen.

Many external sources generate free radicals. These sources include pollutants, cigarette smoke, alcohol, heavy metals (such as cadmium and lead), sodium nitrate (a preservative that is found in cold cuts), chlorine, aldehydes, sulfur dioxide, ozone, and radiation (such as X-rays, overexposure to the sun, cosmic radiation, nuclear byproducts, and fallout). Other toxic free radicals can be generated by cooked or rancid foods, mutagens, carcinogens, heated proteins, mold, toxic plant substances, exhaustion, illness, and perhaps stress.

Toxic free radicals can damage body proteins that become cross-linked and tangled; tissues can lose their suppleness; arteries can harden; and the susceptibility to cancer can increase.

The common dread of cancer has produced an allotment of large sums of money for research on cancer drugs. Because public funds are allocated by politicians, and because many politicians are elderly, it is not surprising that they should sponsor research that could benefit them as well as their constituents. The slow advances of chemotherapy have shown that money alone cannot buy knowledge. Creative ideas, intuition, and correlation of facts remain the bases of progress in any complex study, such as the chemotherapy of tumors. Moreover, scientists from different fields who are not familiar with each others' methods have to learn to work together. These matters take much time and mental effort.

Only some of the drugs now in wide use to treat human malignancies will be mentioned in this section. However, the reader should be aware that more than 500,000 compounds have been tested in animals, mostly in mice. Five different mouse tumors are commonly studied. Experience has shown that 95% of those compounds that are active against all five mouse tumors are also active against human cancers. Likewise, 95% of all drugs effective for human cancers are effective for the mouse tumors. Only a small number of drugs active in mice have advanced to clinical trials, however, because their greater toxicity in humans makes it impossible to try them clinically. Obviously, metabolism differs in humans and mice.

The earliest effective antitumor substances were alkylating agents, and some of them are still in wide use, principally to control leukemias and tumors of the lymph system, as well as some virus-caused cancers. They were discovered during World War II.

An Allied supply ship carrying war gas, sulfur mustard, was landing in the harbor of Naples, Italy, to unload its cargo, in case the German General Staff should, in despair, loosen war gases against American troops. The vessel was bombed, and sulfur mustard spread over the waves. The ship was abandoned, and its occupants were forced into the sea covered with the war gas. Examination of the blood of those who were rescued showed a sharp drop in the number of white blood cells (*leukocytes*). Alfred Gilman Sr. compared this observation with the overabundance of white cells seen in leukemias, but he could not apply sulfur mustard to leukemia patients because of its poisonous effects. Mustard gas is an aggressive chemical that reacts with the proteins and nucleic acids of body cells by irreversibly tying their strands together. Chemists soon devised other compounds with the same binding, or *alkylating*, quality but with less toxicity. The first compounds were nitrogen mustards, which no longer blistered tissues as the sulfur compounds did. The most successful of these drugs has been cyclophosphamide. Still, like the earlier drugs, it is not as specific for cancer cells as one would hope, and it also suppresses protective immunological reactions against foreign cells.

Cancer cells, unlike normal body cells, grow in an uncontrolled fashion. They increase to become tissues that push aside surrounding tissues and invade and damage them beyond repair. Often these malignant cells detach themselves from the original, or primary site, float away in circulating fluids, and lodge or fasten themselves in remote places, where they continue to multiply. This spreading process is called *metastasis*. For example, a thyroid cancer cell may metastasize to some internal organ or a limb and become a second thyroid cancer in that distant spot. Of the more than 100 recognized types of cancer, some grow slowly and others grow fast, and the fast growers metastasize most vigorously.

Some cancers depend for their multiplication on hormones. Breast cancer is accelerated by estrogens; testicular and prostatic cancer, by male sex hormones. Therefore, surgical removal of the ovaries was used commonly to retard the growth of breast cancers. Now, though, antiestrogens, such as tamoxifen, are tried before surgery. Steroidal antiandrogens, such as cyproterone acetate, and nonsteroidal antiandrogens, such as flutamide, are used to combat the growth of cancer in the prostate gland.

A number of human and animal cancers are attributed to viruses. These infectious particles invade the host cells and force upon those cells their own metabolic mechanisms, especially the manufacture of the viral nucleic acids by which they multiply. In some cases, this takeover may lead to continued virus infections; in others, to malignant cell growth.

A virus causing leukemia of human T cells (thymus cells) has been identified with the virus that produces acquired immune deficiency syndrome (AIDS). AIDS is a relatively new disease that suppresses immunological control of infections and of certain cancers. No therapy is known to cure AIDS, but a few drugs originally designed as antitumor agents appear to delay some of the symptoms of the disease. Among them are azidothymidine (AZT), and a number of related agents that act as antimetabolites.

Some medicinal chemists dream of finding a way to block the oncogene portion of a body cell's reproductive machinery so that a virus or some chemical cannot trigger it into causing the mutation that leads to uncontrolled cell growth. All genes are segments of nucleic acids—DNA or RNA—which are very long molecules. Some of these nucleic acids are coiled in a double spiral, or helix, but they must uncoil to give directions for copying each half when the time comes for a cell to divide or to give directions for making proteins.

Scientists now know the general pattern of the nucleotide segments, the types of carbohydrates and bases in them, and the nature of the links that fasten the segments together, and they can understand the meaning of one of the segments that will change the directions it gives for making a metabolic substance (*metabolite*). The new metabolite is likely to act as a metabolite antagonist that interferes with the original substance's production or reactions. Should some chemist modify the oncogene's nucleic acids so as to produce a suitable nucleotide metabolite antagonist, then cancer might be stopped.

However, we must consider the mathematics of the chances. In each nucleotide segment there can be one of four bases and one of two carbohydrates. If only the hydrogen atoms in them are counted, each carbohydrate has about six and each base has four or five, and there are also two to four nitrogen atoms that can be replaced or shifted around. Add these numbers, and observe that the possible variations would result in tens of millions of potential metabolite antagonists. It might take half a year to make one, and it would take six to eight years to train someone to make them. All told, it would take several generations to cover this ground even if everybody were drafted for the task, not to mention testing the compounds after they were made. Who would be left to raise food or sell automobiles? That is the problem of cancer chemotherapy. For four decades, dedicated chemists have synthesized thousands of metabolite substitutes, and equally dedicated biologists have tested them. A few of these metabolite substitutes are suitable to test in the clinic, but these are just a drop in the bucket compared to what it would take to find effective drugs against all cancers. It is almost a miracle that any have been found at all.

Some of the new drugs block enzymes that put the segments together to form the long nucleic acids. Other drugs enter into a segment themselves but do not fit the code and so prevent the nucleic acids from functioning. Examples are two widely used antimetabolites: 6-mercaptopurine and 5-fluorouracil.

Whether an active drug will be found is always a gamble, and compounds with no relation to known effective substances are not likely to be studied. Nevertheless, some totally unexpected chemicals have turned out to be effective against some cancers. The best known of these are antibiotics. The original antibiotics were organic compounds obtained from fermentation products of microbes, some of which can retard the multiplication of disease-causing microorganisms, parasites, or cancer cells. The first antitumor antibiotics came from the *Actinomyces* mold, but the only potent, useful one is actinomycin D, which is still too toxic. Another group of antibiotics active against several human cancers are the bleomycins, originally fermented by *Streptomyces verticillatus*. Doxorubicin (adriamycin) is obtained by fermentation from *Streptomyces peucetius*. The last two have been synthesized. These substances bind to DNA and thus prevent it from uncoiling to make the RNA directions for protein synthesis and even to begin cell division.

A completely different substance with anticancer activity is the inorganic compound *cis*-platinum (*cis*-dichlorodiammineplatinum(II)). This agent, never suspected of having biological importance, was tested in 1967 against intestinal bacteria and later as an antitumor drug. It may act similarly to the alkylating agents. Although it is toxic, its value in cancer treatment is undisputed.

Among the natural alkaloids tested against rodent tumors, those from the periwinkle plant (*Vinca*) are especially useful in childhood leukemias. They are vinblastine and vincristine. A chemically complicated compound called taxol has been isolated from species of the yew tree. It is effective against breast cancers.

A cancer specialist cannot always determine the biochemical reactions to which the cancer might be most sensitive. Therefore, cancer treatment rests on the practical assumption that multiple attacks by various drugs are more likely to hit the tumor. This is why pharmaceutical firms usually mix three or more drugs in one tablet, capsule, or injectable preparation. The most common combinations are one alkylating agent, one antimetabolite, and one antibiotic (bleomycin or adriamycin), with *cis*-platinum, vincristine, or other substances added as needed, depending on the type of cancer. Such shotgun treatment has produced some cures and raised the five-year rate of remission for some malignancies to 50–75%.

Drugs for the chemotherapy of cancers almost invariably cause objectionable side effects such as loss of hair and weight loss. Other nonchemical methods of halting metastases may be preferable to chemotherapy in some cases, such as carefully directed irradiation of malignant tumors and surgical removal of tumor tissues. The spread of cancer cells during surgery can be minimized by concomitant chemotherapy. Until researchers find more specific drug therapies to combat cancer, doctors will continue to use the three-pronged attack method: surgery, drugs, and radiation.

15

Drugs Affecting the Immune Response

For centuries, survivors of some diseases, such as smallpox, have been known to be immune to a second infection. The first infection by viruses or other pathogens produces a chemical toxic to the host, called an *antigen*. After 10–20 days, some new proteins are built that are patterned on the antigen. These new proteins are called *antibodies*, and they neutralize the effect of any new dose of antigen that may enter the body on a subsequent occasion. The acquired immunity may last for a short while or for a lifetime. Immunization with specially prepared antibodies may thus need to be given only once, as in the case of poliomyelitis, or at regular intervals, as with immunizations for smallpox, typhoid fever, cholera, yellow fever, and influenza.

The second type of immune response is mediated by cells of the thymus gland (T cells) and the B cells, named for the bone marrow. These responses cause delayed hypersensitivity, as encountered in allergies, and in some cases protection against infection. The most serious consequence of cell-mediated immune response is rejection of foreign materials, such as grafts and organ implants. Suppression of this response can lead to some acceptance of transplanted organs and resistance against some diseases that produce cells or tissues rejected by the body as if they were foreign to the organism. A few such diseases are psoriasis, myasthenia gravis, multiple sclerosis, rheumatoid arthritis, lupus erythematosus, rheumatic heart disease, and perhaps early-onset diabetes.

Besides these diseases, at least 100 others involve cell-mediated immunity. It is now believed that many, if not all, of these diseases, are caused by the lack of a normal gene, or by the presence of mutated or faulty genes in a genetically predisposed individual. Thus, such diseases are inheritable according to the laws of evolution.

Genes are segments of nucleic acids; they consist of a given sequence of nucleotides, the building blocks of the nucleic acids. In some genes this sequence is one piece of the polynucleotide chain, whereas other genes consist of two such sequences, separated from each other by sections of the nucleic acid unimportant to the activity of the gene.

RESEARCH INTO THE IMMUNE RESPONSE

We do not know yet what approaches will be valid to inhibit faulty genes and stop them from exerting their disease-causing influence. Because the genes arrange the way in which the amino acids in the circulation should combine to yield a certain protein, the protein resulting from obeying these commands from faulty genes contains faulty sequences of amino acids. At present, it seems most feasible to try to prevent the biosynthesis of such faulty proteins by drugs that look a little like normal peptides (combinations of several amino acids) but cannot take the place of the normal peptides and therefore block the assembling of amino acids to form faulty proteins. Such drugs that imitate ordinary proteins are called *peptidomimetics*. Some of them have already gained recognition as specific enzyme inhibitors and receptor agonists and antagonists and have been tested for curative activity in genetically caused diseases, including some types of cancers.

One approach to achieve this curative activity is to let a virus produce a normal gene and inject the altered virus into an animal suffering from a genetic aberration. The normal gene will replace the faulty gene of the animal and thereby cure it. This procedure has been performed with an adenovirus in mice that had been previously infected with a muscular dystrophy gene.

Another genetically engineered enzyme is the human enzyme, deoxyribonuclease (DNase), which breaks down DNA. This nucleic acid is found in excessive amounts in the secretion of the lungs of cystic fibrosis patients. It thickens the mucus, which contains dead white blood cells that accumulate to fight infections. For relief from this situation, DNase is inhaled through a nebulizer. The Dnase disperses the thick lung secretions and thereby decreases the frequent bacterial infections that threaten lung tissues. DNase is not a cure for cystic fibrosis, but it improves lung function and the quality of life in such patients.

One type of cell arising from normal body cells by mutation and then becoming "foreign" to the body is the cancer cell. When cancer chemotherapy became a clinical reality, virtually all anticancer drugs were also tried to suppress cell-mediated immune responses. The results have been variable; at best, these drugs did their job for only a limited time and then lost their suppressive activity. Generations of organ-transplant patients and patients with autoimmune diseases have had to contend with this diminishing suppression. Changing the drugs sometimes extends the period of suppression, but only seldom could this suppression be maintained for a prolonged time.

In the case of methotrexate, a typical sequence of events was seen. The drug was active, but only at high doses that caused considerable toxicity. Because methotrexate is a dihydrofolic acid antagonist, dihydrofolic acid was administered to patients simultaneously. The dihydrofolic acid counteracted the toxicity of methotrexate without completely ruining its antiimmune response effect—at least for some time.

Among the agents tested were all the antitumor agents, including some antibiotics. A virtual revolution in immunosuppressive therapy occurred when one of these antibiotics, cyclosporin A, was tried. Cyclosporin A is a peptide consisting of 11 amino acids arranged in a large ring shape. It was discovered by a Swiss research team in 1976 by fermenting the fungi *Cylindrocarpon lucidum* and *Trichoderma polysporum*. Its peculiar molecular composition makes it resistant to being digested; therefore, the drug can be taken orally. It is much more effective in suppressing cell-mediated immune responses than any other agent used previously.

An unexpected application of cyclosporin A is in the treatment of early-onset diabetes. It now appears likely that childhood diabetes may be caused by an abnormal genetic tendency of the child's T cells to destroy the beta cells of the pancreatic islets of Langerhans that produce insulin. Administration of cyclosporin A to such individuals, even without the customary doses of maintenance insulin, has given favorable results in suppressing the symptoms of the disease.

BIOTECHNOLOGY

On the preceding pages, several drugs have been listed that are based on genetic engineering. Human insulin is produced by microbes that have been altered genetically; DNase is used in cases of cystic fibrosis, and human growth hormone is produced commercially. Genetic engineering is part of the wider field of biotechnology, in which biological organisms and enzymes from natural sources are used in place of conventional chemicals. All these efforts are in turn based on molecular biology, which deals with the study and the uses of biological preparations in chemical terms.

The earliest uses of biotechnology in our sense of the term can be discerned in fermentation processes. The first fermentation used the carbohydrates contained in grains to make ethyl alcohol by means of enzymes from yeast (*Saccharomyces cerevisiae*) and from hops, the dried ripe cones of the female flowers of *Humulus lupulus*. Some other industrial chemicals are also produced by the fermentation of carbohydrates, such as citric acid. The most intense use of fermentation is producing antibiotics through large-scale metabolic reactions of molds, bacteria, and other microbes.

Other microbial reactions have been applied to shorten or facilitate synthetic sequences and thereby reduce the cost of the end product. For example, the early commercial synthesis of cortisone required more than 30 discrete chemical operations; one fermentation procedure shortened this synthesis by eight steps.

Biotechnology has not only simplified chemical reactions but has made possible access to hitherto unobtainable materials. Thus, some steroidal derivatives as well as peptides and proteins that had withstood chemical synthesis can be produced by enzyme-catalyzed metabolic reactions.

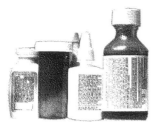

16

Drugs for Infectious Diseases

Everything that lives, metabolizes, and reproduces depends on the chemicals in other living or dead cells and tissues for survival. Herbivorous animals feed on plants, carnivorous animals feed on animal tissues, and omnivorous animals feed on both plants and animals. Snakes devour living rodents; tigers kill their prey before feeding on them. Humans take a middle road, using vegetables, fruit, live animals (such as marine shell organisms), and dead animals for their diets. Seabirds kill their own kind before devouring them, but they eat live fish that are still flapping around while being ingested. Cannibalistic humans are still to be found more often than one wants to admit.

Microbes are organisms at the bottom of the plant and animal kingdoms, but their dietary requirements are not much different from those of higher animals or carnivorous plants. Microbes need the building blocks of proteins, nucleic acids, many hormones, some vitamins, fats, and carbohydrates to survive, to reproduce, and to supply the energy for their life processes. Some microbes invade foreign organisms and live with them in cooperative arrangements. Many symbiotic bacteria inhabit the intestinal tracts of mammals, feast on the diets of their hosts, and furnish the hosts in return with essential biocatalysts, such as certain vitamins that the bacteria can synthesize while the hosts cannot. Some of these bacteria digest cellulose for their hosts, converting it to glucose that the hosts (such as ruminant animals and termites) can then use.

In many other cases, a microbial invasion of a human or beast takes its toll on the host organism. Infectious organisms not only use the host's biochemicals for their own nutrition but usually metabolize them to materials that are toxic to the host animal and, in sufficient amounts, kill it. In a way, this is a shortsighted action in Nature's household. After the death of the host animal, the flow of its biochemicals stops, and the invading microbes die from the lack of needed nutrients.

When people talk about infections, they usually think of the many infectious diseases that befall humans, farm animals, pets, and wildlife. But infections are also common among other forms of life. Viruslike particles (*bacteriophages*) assault bacteria, and some bacteria dislodge others from their source of nutrition. An endless chain of predatory events threatens every living cell and even subcellular particles in a steady fight for survival. On the highest level, humans destroy food animals, and they destroy other humans by warfare and other modern methods. There is no mercy in nature. Some microbes can be kept from reproducing by the use of some chemical weapons against them. Such drugs are called *bacteriostatic* or *virustatic*. If a drug *kills* the invading microbe, it is called *bactericidal* or *virucidal*.

A new weapon used in warfare usually takes a heavy toll on the unprepared victim of the attack. After the surprise has worn off, the victim often finds ways to avoid the same kind of attack on future occasions. Likewise, infectious microbes often succumb to a new foreign chemical that inhibits some of the enzyme catalysts essential for their life processes. After a number of these experiences, and with a continued indomitable destiny to survive, microbes search for means to avoid the inhibition of their vital enzyme systems. Their metabolism invents reactions that do not need these enzymes. These shortcuts or circumventions are called *metabolic shunts*. When such shunts become established, the microbe is no longer affected by attacks on its previously essential enzymes. It has become resistant to the drug.

All chemotherapeutic drugs have in common a fundamental mechanism of action. They are directed against the life processes of the invasive, pathogenic organism. Drugs stop or interfere with the biosynthesis or usage of the building blocks of the invader. These building blocks can be constituents of the cell membrane or of the cell nucleus of the pathogenic cell. Usually the action of drugs is directed against protein synthesis, be it the biosynthesis of one of the amino acids that make up the proteins, or the assembly of the amino acids into a large protein molecule. Other drugs interrupt the biosynthesis of the constituents of nucleic acids; they can also interfere with the assembly of these pieces into nucleic acids, or with the usage of the nucleic acids to transmit commands for protein synthesis. All this activity goes on in the chemotherapy of all invading cells or organisms, be they bacteria, viruses, amoebas or other protozoa, fungi, pathogenic yeasts, or malignant tumor cells. Chemotherapy is essentially a strategic blockade, aimed at the slow death or destruction of the cells to be halted. In most cases, the defenses of the animal body, the *phagocytic* blood cells, try to overcome the infectious invaders by enclosing and digesting them, but these defense cells cannot overcome the ever-increasing number of the attacking pathogens. That is where chemotherapy comes in. Drugs either kill the pathogens or make it impossible for them to reproduce. The body's defense cells then take care of the smaller enemy armies.

CLASSIFICATION OF PATHOGENIC PARASITES

Disease-causing (*pathogenic*) foreign organisms range from minute bodies without membranes to multicellular animals that invade host animals. Only a few of these organisms will be catalogued here, and only those that cause the more common infectious diseases observed in moderate climatic latitudes. Some of these infectious organisms belong in the plant kingdom. Others are low forms of animal life. Some pathogens, such as the viruses, have affinities to plant or animal cells but are neither themselves.

Viruses

Viruses are the simplest forms of life. They are microscopic assemblies of genetic material, either DNA viruses or RNA viruses. The nucleic acids are encapsulated in protein in a *capsid* that sometimes also contains carbohydrate and lipid components. Viruses can infect plants, insects, bacteria, or animals. They depend on the host cells for nutrients and energy. Animal viruses—the most important in medicine—attach themselves to the surface of the host cells and penetrate into the cell. The protein capsid detaches itself from the nucleic acid materials, and this bared genetic substance begins to direct the host cell to synthesize the proteins and enzymes needed by the virus. These new substances, manufactured by the subjugated host cell at the command of the invader, are assembled to form a new virus. The new virus is then extruded and can infect a new host cell. The few chemotherapeutic drugs found that are effective in viral diseases interfere at some of these stages of viral life processes. A few of the hundreds of viral diseases are the common cold, poliomyelitis, viral encephalitis, some forms of leukemias and other cancers, herpes, influenza, yellow fever, acute respiratory disease, smallpox, fowl plague, and AIDS.

Rickettsiae

Rickettsiae have rudimentary cell walls and resemble bacteria more than they resemble viruses. Like bacteria, they can synthesize some proteins, and like viruses, they live mostly inside host cells. Among diseases caused by rickettsiae are psittacosis (parrot fever) and lymphogranuloma venereum. Rickettsias (diseases caused by rickettsiae) are transmitted to humans by tick bites and produce such infections as Rocky Mountain spotted fever. Another infection transmitted by the tick *Ixodes scapularis* is Lyme disease, which is named for Lyme, Connecticut, where it was first observed; the victim receives a bacterial spirochete, *Borrelia burgdorferi*, through the tick bite, and this spiral microbe then sets in motion the disease.

Other Intermediary Pathogens

Several other pathogens have characteristics between those of viruses and those of bacteria. Some of them cause dangerous human diseases. The differences between them are largely their gradual independence from host cells, some sophisticated nutrient requirements, and thereby their susceptibility to drugs that interfere with some new life processes. Among these organisms is a mycoplasma that causes primary atypical pneumonia (also known as viral pneumonia). L-forms, spheroplasts, and protoplasts are modified forms of bacteria. They are extruded by some bacteria or revert sometimes to bacterial forms. They have cell walls, although sometimes defective, are resistant to penicillin and similar antibiotics, and contribute to difficulties in subduing some bacterial infections.

Bacteria

The best studied unicellular organisms enclosed in a rigid cell wall are the bacteria. The bacterial cell wall consists of polysaccharides (celluloselike materials) and proteins that have no analogy in animal membranes. Thus, chemotherapy can concentrate on attacking processes that build the bacterial cell wall because such interferences cannot damage cell wall synthesis in the host animal.

Bacteria are free-living organisms that can synthesize their metabolites from simple compounds, including inorganic salts, as long as they contain the essential elements of all living matter. Bacteria can thrive when they are independent of host cells, but in some cases their nutritional requirements have to be produced by host organs. Some bacteria need an oxidizing environment (*aerobic* bacteria), whereas others get along better under nonoxidizing (*anaerobic*) conditions. Resistant strains of bacteria are the dread of chemotherapists. They emerge through mutation, and drugs can lead to their formation if the drugs stimulate bacterial cells to learn to synthesize just those enzymes that the drug inhibits. In other words, the bacterial cell learns to defend itself until the drug loses all its therapeutic value.

Protozoa

Bacterial cells belong substantially to low forms of plant life. The lowest forms of animal cells comprise many types that are pathogenic to mammals, including humans. They belong to the zoological classification of the phylum Protozoa (derived from *protos,* meaning first and *zoion,* meaning animal), one-celled animals that cause a variety of lethal diseases.

Amebiasis is due to *Entamoeba histolytica,* a parasitic protozoan species of amoeba with a complicated life cycle. Trypanosomiases are diseases caused by unicellular protozoan parasites of the genus *Trypanosoma.* Humans and animals alike become infected with trypanosomes through the bite of flies, for example, tsetse flies. In Africa, the human infection is called sleeping sickness; in South and Central America, Chagas' disease. It can no longer be presumed that such infections are restricted to tropical areas of the earth. The rapid exchange of travelers as well as of animals and the return of military and industrial personnel from countries where such diseases are endemic has exposed the northern hemisphere to outbreaks of a number of so-called tropical diseases.

Malaria is such a case. This group of diseases is transmitted by protozoan organisms of the genus *Plasmodium*. The most dangerous species for humans is *Plasmodium falciparum*, but other forms of malaria caused by *P. vivax* (benign tertian malaria), and *P. malariae* (quartan malaria) also cause much damage, particularly because the causative plasmodia are recycled, passing through a multistage life cycle. Plasmodia spend one stage of their lives in anopheles mosquitoes. Infected female mosquitoes inject the earliest life form of plasmodia, the *sporozoites*, into humans or animals when the mosquitoes feed on the blood of their victims.

Chemotherapy made inroads into protozoan infections long before bacterial infections yielded to drugs. Most viral diseases cannot yet be treated with reliable and satisfactory drugs; their state of chemotherapy is where that of protozoan diseases was 80 years ago. However, several enzymes essential to the viral life processes are now understood, for example, proteases, which play a critical role in the biosynthesis of viral protein coats. New experimental drugs can be tested as inhibitors of these enzymes and thereby speed up the process of discovery and selection of useful agents. In other words, one can now design potential drugs more intelligently than could be done with other chemicals at the dawn of the chemotherapeutic era.

CHEMOTHERAPY OF BACTERIAL INFECTIONS

By the early 1900s, hundreds of attempts had been made to protect patients from bacterial infections, but none of the chemicals tried had fulfilled the two cardinal requirements: high efficiency coupled with low toxicity. Many of the old agents were efficient but much too toxic. Others that had acceptably low toxicity were not effective against a sufficiently wide variety of bacteria.

True chemotherapeutic drugs for bacterial infections were first conceived and explored by two great pharmacologists, Gerhard Domagk (1895–1964) and Daniel Bovet (1907–   ). The story of their discovery sheds light on inventiveness and motivation in drug research.

Sulfanilamide, the original prototype of these drugs, was prepared by a German graduate student in 1907, just out of chemical curiosity. It was then forgotten. It is a simple organic compound, and generations of premedical undergraduates in this country have prepared it as a synthetic exercise during their year in the organic chemical laboratory. In the meantime, Paul Ehrlich (1854–1915) suggested that dyestuffs that could stain pathogenic microbes selectively on a microscope slide might kill the same microbes without affecting the surrounding host tissue if they could be made selectively toxic for the pathogens. Ehrlich called this visionary concept chemotherapy, a term he coined for this purpose. Inspired by his vision, every dyestuff—tens of thousands of them—in the archives of the dyestuff industry was tried against bacterial cultures in glass dishes and test tubes, but to no avail. The dyestuffs did not stop the multiplication of the bacteria.

Scientists get fed up with doing test after test, each the same way, with nothing as a guide. A number of dyestuffs on wool fabrics had proven to be resistant to sunlight and to washing. Wool is an animal fiber and consists of fibrous proteins in contrast to cotton, which consists of the polysaccharide cellulose. If a dyestuff adheres to a protein fiber, Ehrlich reasoned, will it not stain other proteins as well—say, bacterial cells? He tried these wool dyes on the bacterial cultures in their glass containers based on that hunch, but again they did not work.

At this point, the German pharmacologist Gerhard Domagk suggested that the wool dyestuffs be tried in bacteria-infected mice. After all, chemotherapy was supposed to protect animals from infections or to heal the infection. And then the miracle happened: Mice infected with deadly staphylococci were injected with solutions of the azo dyestuffs, and they recovered—not all of them, but most of them. Peculiarly, only those dyestuffs worked that contained a chemical group called sulfonamide, —SO_2NH_2. Encouraged by this breakthrough, Domagk tried the least toxic of the sulfonamide dyes in patients with staphylococcus infections. So confident was he that he chose his own daughter, Hildegard, as the first patient for these trials. She recovered, and her father was offered the Nobel Prize in medicine. The registered trade name of the dyestuff was Prontosil. It was marketed all over the world and began to save the lives of patients with heretofore dreaded infections, such as hemolytic streptococci and staphylococci. In this country, the son of President Roosevelt, Franklin Delano Roosevelt, Jr., was saved by that drug.

Prontosil (also called Prontosil rubrum because of its red color) became a profitable drug for the German Bayer Company, which held the patent rights. To protect their treasured property, they did not publish the chemical structure of the dyestuff, but it did not take long to break that secrecy. Chemists at the Pasteur Institute of Paris did some detective work, scanning the patents of Bayer Company, and ultimately reconstructed the chemistry of Prontosil. They made many similar dyestuffs in this process to support their conclusions.

In 1935, the pharmacologist Daniel Bovet at the Pasteur Institute kept a colony of mice infected with pathogenic cocci and tested the dyestuffs for curative effects. Something strange happened to the dyestuffs, though. The urine of the mice was colorless, not red or any other dyestuff color. The mice had metabolized the dye to a colorless compound. This compound was soon identified. It was a simple derivative of sulfanilamide, which had been synthesized by that graduate student 28 years earlier. A bottle of commercial sulfanilamide, a white powdery substance, stood on the shelf at the Pasteur Institute. Indeed, it was used every day to make all those dyestuffs. Bovet tested sulfanilamide and saw that this colorless, nondyestuff drug cured the infected mice. Did the German company observe that cure also? And if they did, did they keep silent about it because sulfanilamide was an old compound and not protected by a patent? All the people involved in the discovery of Prontosil are gone, and we will never know the truth.

The antibacterial potency of sulfanilamide varied in several in vitro experiments. The British biologists P. Woods and D. D. Fildes treated sulfanilamide-sensitive bacteria with various nutrients in glass dishes and noted that without beef broth present, the bacteria were killed by the drug. With the broth, they were kept from multiplying, that is, sulfanilamide was only bacteriostatic. They analyzed the broth and found that it contained *para*-aminobenzoic acid (PABA), an essential growth factor for the bacteria. Woods and Fildes had the presence of mind to compare the chemical formulas of sulfanilamide and PABA. These simple formulas are reproduced here because of their obvious similarity: PABA is $H_2N-C_6H_4-CO_2H$, and sulfanilamide is $H_2N-C_6H_4-SO_2NH_2$. We have encountered structurally similar compounds with opposite biological actions before. In this case, PABA and sulfanilamide were recognized as the first important pair of metabolic antagonists.

The question still remained why the bacteria needed PABA. It was found that a bacterial enzyme helped PABA to be incorporated in the molecule of the B vitamin, folic acid. Reduced folic acid, in turn, is needed for the synthesis of the purine bases present in nucleic acids. The antagonism of sulfanilamide to the incorporation of PABA into dihydrofolic acid thus deprives the bacteria of the materials in their cell nuclei and thereby of life. Some bacteria that cannot use PABA to synthesize dihydrofolic acid use prefabricated folic acid from the cells of their hosts. Such bacteria are not affected by sulfanilamide; they are resistant to the drug.

Sulfanilamide itself was too toxic for extended routine treatment of infections with cocci. The usual procedure of molecular modification created a long line of analogues with appreciably fewer side effects and with greater potency and longer duration of action. The best analogues also avoided a real drawback, namely, crystallization of drug metabolites in the kidney tubules. Among the analogues that have remained of use in chemotherapy are sulfathiazole, sulfadiazine, sulfisoxazole, and sulfasuxidine. Each of them has its own set of advantages. Sulfisoxazole is well suited to reach bacterial pathogens in the kidney and bladder. Sulfasuxidine removes bacteria from the intestines, an action important for bacterial diarrheas and for cleaning up the intestines before intestinal surgery.

Sulfanilamide and its analogues deny PABA access to an enzyme that helps folic acid synthesis. Later on, the folic acid must be reduced by another enzyme, dihydrofolate reductase, before it can participate in the construction of materials needed in the bacterial cell nucleus. Several drugs unrelated to sulfanilamides inhibit the enzyme dihydrofolate reductase and thereby block all the subsequent steps in this long cascade of biochemical events. If the sulfanilamides block one step and the dihydrofolate reductase inhibitors block another step, would a combination of these two types of drugs not be even better? This combination was tried by George H. Hitchings, and it worked. A combination of trimethoprim (a dihydrofolate reductase inhibitor) and sulfamethoxazole has become one of the most effective antibacterial agents.

Some patients feel that it is a nuisance to have to take a drug several times a day or, in extreme cases, to have to be wakened at night so that they will not miss a dose. Several sulfanilamide derivatives have been found that have half-lives of 65 hours and even 150 hours (meaning that only one-half of the drug has been excreted after that time). However, these attractively long-lasting agents have some drawbacks that have not been completely overcome.

A relative of the sulfonamide drugs, called dapsone, has become the most effective agent for the treatment of leprosy. This ancient and feared mycobacterial infection has surfaced in developed as well as underdeveloped regions. Except for a few virtual reservations in which patients live together, the northern hemisphere is free from this slow-onset infection.

The much more prevalent mycobacterial disease worldwide is tuberculosis. This infection also begins slowly, but it is difficult to treat chemotherapeutically. An estimated 50,000 chemicals have been tried against tuberculosis in laboratory animals. The first drug to show promise was *para*-aminosalicylic acid (PAS), but PAS is metabolized quickly, and large doses have to be given orally to maintain its concentration in the blood. The mycobacteria also have a tendency to become resistant to PAS.

The sulfanilamides are, on the whole, ineffective against tuberculosis, but the weak activity of one of them led Domagk to study some related compounds and to single out some of the synthetic intermediates needed in the preparation of the substances he wanted to test. Some of these intermediates, called thiosemicarbazones, had a modicum of activity, and the best of them was to be made in a larger quantity for further tests. These experiments were carried out in the pharmaceutical industry, where time is precious. It took a long time to get these syntheses going, and the chemists felt hard-pressed to supply samples to the biologists for tests, so they gave them every chemical they had, including, again, some intermediates. And one of these scored: It was isoniazid, which in 1951 rapidly became the most important antituberculosis drug. It was followed later by ethionamide and pyrazinamide, both industrial products of molecular modification.

These vast efforts involving hundreds of medicinal scientists took place about 1950, when some antibiotics had already made their appearance. Streptomycin and dihydrostreptomycin, introduced by Selma Waksman, were the first antibiotics with activity against tuberculosis. They have been largely abandoned because of their toxicity on the auditory nerves. The antibiotic that has been of great benefit in chronic and drug-resistant pulmonary tuberculosis is rifampin, a semisynthetic derivative of the rifamycin antibiotics fermented by *Streptomyces mediterranei*. Its very difficult chemical and biological study was performed by Piero Sensi in an Italian pharmaceutical firm.

Tubercle bacilli, like any other infectious organisms, tend to mutate, especially when they are constantly confronted with lethal chemotherapeutic agents. Such mutated tubercle bacilli have become a new danger in recent years. They are no longer susceptible to chemotherapy by the accustomed drugs. A whole new effort to overcome this newly emerged resistance has to be mounted, inventing new, "second generation" drugs that the bacteria have not previously encountered.

Bacteria and other microbes are grouped into two large classes, gram-positive and gram-negative. This classification is based on their ability to retain a dyestuff stain originated by Hans C. J. Gram (1853–1938). Gram-positive organisms are, on the whole, more easily affected by drugs than gram-negative ones. There are still many gram-negative bacteria for which no workable drug has been developed. This is true even of antibiotics, although modified penicillins, cephalosporins, and other drugs can be used in some gram-negative infections.

ANTIBIOTICS

Antibiotics are chemicals that interfere with the life processes of pathogenic microorganisms, prevent their multiplication by antagonizing their reproduction, and occasionally destroy the pathogens. The term *antibiotic* was coined by Waksman in 1942 to designate antimicrobial substances produced by other microorganisms. This definition is too narrow and is no longer accepted. Many antimicrobial chemicals are used systemically that are and always have been purely synthetic, for example, the sulfanilamides. Several natural antibiotics obtained by the fermentation of microbes have been synthesized and are now manufactured synthetically. Many plant products exert antibiotic actions indistinguishable from those of classical fermentation antibiotics. And last but not least, several classical antibiotics kill cancer cells and parasitic multicellular animals, such as pathogenic worms, that cannot be called microbes.

History of Antibiotics

The treatment of infections and infestations with what we would now call antibiotics has been known for thousands of years. Ancient Chinese savants treated boils, carbuncles, and other skin infections with moldy soybean decoctions, the mold serving as a source of an antibiotic. Louis Pasteur and Joubert found in 1877 that "common bacteria" could kill cultures of anthrax bacilli and act similarly in anthrax-infected guinea pigs. An impure material exuded by *Pseudomonas aeruginosa* proved to have some effect against diphtheria as early as 1890. We now know that the material contained two substances: an enzyme called pyocyanase and an antibiotic called hemipyocyanine.

Naming of Antibiotics

Most names of antibiotics reflect the name of the genus of the micro-organism from which they are produced. Thus, penicillin comes from *Penicillium*, polymyxin from *Bacillus polymyxa*. Others have been named for the country or place where the original source, a soil sample, was collected. Angolamycin is derived from Angola, foromacidin from the Roman forum, miamycin from Miami. Nystatin relates to the New York State Board of Health Laboratories and hamycin to Hindustan Antibiotics, Ltd. Wives of the investigators have furnished some names, such as doricin and helenin. Supportive and indispensable secretaries have been the source of other names such as nanimycin and vernamycin. Even a mother-in-law's name was drafted for this purpose: seramycin. A girl by the name of Tracey suffered from a bacillary infection that was cultured and provided the source of bacitracin. Rifamycin was named after the motion picture *Rififi*. Some of these names were altered later by the firms that supplied these drugs.

The Penicillins

Penicillin was discovered by the British bacteriologist Alexander Fleming in September 1928. A culture of staphylococci growing on an agar medium became contaminated by a mold that had been carried to it in the laboratory air from some distance. The bacteria growing in the vicinity of the mold disappeared. Fleming identified the mold as *Penicillium notatum* and realized that it must have produced an antibiotic substance, which he named penicillin. He showed also that this fermentation product did not make the culture medium useful as an antiseptic. The day of systemically active antibacterials had not yet arrived (Prontosil surfaced in 1932), and it never occurred to Fleming to explore any systemic action of penicillin.

Not until 1937 was the real, useful antibiotic elaborated by a team of chemists and biologists at Oxford University (H. W. Florey, E. B. Chain, E. P. Abraham, and N. G. Heatley). The first clinical use of penicillin was made on February 12, 1941, on an Oxford police officer with a septic infection. For several years, so little penicillin was available that any unchanged antibiotic had to be recovered from the urine of the patients and repurified for the next injection.

The term *penicillin* covers many drugs. Early in its investigation, more than seven similar but chemically different penicillins were found among the fermentation products. Because the chemistry of these penicillins emerged only slowly, a great deal of confusion accompanied the study of "penicillin." Clinically, these products were not equivalent either, and finally one of them, benzylpenicillin, was chosen as the most useful of these natural penicillins. It has become "the" original penicillin of medicine. Benzylpenicillin must be given parenterally, that is, otherwise than by way of the intestines. This drawback led to the synthesis of orally active analogues that have replaced the natural products of the mold.

Fleming's original *Penicillium notatum* did not give a good yield of fermentation products. In 1944, R. D. Coghill at the Northern Regional Research Laboratories of the U.S. Department of Agriculture in Peoria, Illinois, isolated a strain of *Penicillium chrysogenum* from a moldy grapefruit and got a much higher yield of antibiotic from its fermentation. This strain has become the granddaddy of all the penicillin-producing molds all over the world. Another improvement was the introduction of cornsteep liquor as a nutrient for the fermenting mold. Cornsteep liquor is made by steeping cleaned corn grain in warm water containing 0.2% sulfur dioxide, drawing off the water, and concentrating it under vacuum. It serves as a source of glucose and furnishes 1000 times more penicillin than the older yeast extracts. Also, it became possible to "feed" the mold various chemicals that not only improved yields but led to the isolation of chemically modified penicillins—especially phenoxymethylpenicillin (penicillin V), which has some oral activity.

The differences among the penicillins is in the nature of the small chemical group (the side chain) that is always attached to the same main portion of the penicillin molecule, which is called 6-aminopenicillanic acid (6-APA). This substance became available on an industrial scale in 1959 and opened the floodgates to molecular modification of the penicillins.

Any desired analogue of penicillin could now be made in one simple chemical operation. The pharmaceutical industry lost no time making use of this opportunity. An estimated 10,000 penicillin analogues have been prepared and tested. From this research effort have come the modern penicillins. Some of them are orally active and need not be injected. Others overcame the resistance of bacteria that had sprung up in response to the earlier penicillins. Some of them are active against a number of gram-negative pathogens, others against malarial plasmodia and other infectious organisms. Molecular modification has given the chemotherapist a selection of drugs for various infectious diseases.

All the penicillins have two things in common. First, they are relatively unstable chemicals, some more so than others. That is why the natural penicillins, the least stable ones, had to be injected rather than taken by mouth, because they broke down in the acidic stomach. This disadvantage has been remedied by careful chemical design of the side chains.

The second feature of all penicillins (and cephalosporins and other beta-lactam antibiotics) is their highly selective effect on the bacterial cell wall. Bacterial cells are enclosed in thin cell walls consisting of polysaccharides and proteins in tightly packed, cross-linked structures. The penicillins block the biosynthesis of these cell walls at one little point, the penultimate step of constructing the membrane network. Animal cells have no analogous membranes and are therefore immune to this interference by beta-lactam antibiotics.

The Cephalosporins

After the publication of penicillin studies, every pharmaceutical scientist, executive, and sales representative was given a spatula and a box full of vials and was asked to dig up soil samples wherever they went. Samples were collected in New Jersey backyards, in summer gardens in New England, in forests in national parks, in Venezuela, western Australia, on Alpine slopes—all over. The samples were taken back to the lab, leached out, and worked up for potential antibiotic content. The yield from this aimless gamble was negligibly small, as was to be expected. Moreover, the same old antibiotics were isolated time and again from soils collected thousands of miles apart.

Professor Giuseppe Brotzu of the University of Cagliari on the Italian island of Sardinia collected sewage and grew an antibiotic-producing mold of the *Cephalosporium* species from this unappetizing source. He hit pay dirt: One of the components of the antibiotic broth was less active than benzylpenicillin but two to six times more active against gram-negative organisms. An Oxford group of microbiologists worked up the crude material and discovered that the new antibiotic had some similarities to penicillin, especially the beta-lactam structure, but it was more stable to acids. Molecular modification took over, and total syntheses of the complicated compounds were achieved, especially by the Nobelist Robert Woodward in 1966.

The cephalosporins from this work, which covered more than 15 years, are useful in many previously untreatable infections. A breakthrough occurred when orally active cephalosporins and, later, long-acting cephalosporins were encountered. All of these products are semisynthetic.

An additional medical bonus was realized when the cephalosporins turned out to be much less allergenic than most penicillins. About 19% of all people do not tolerate penicillin. In the early days of antibiotics, penicillin was administered indiscriminately in large doses because it was believed to be nontoxic. It is indeed nontoxic, but its potential for causing severe allergic reactions has led to caution in its use.

Other Antibiotics

Approximately 90 antibiotics are in commercial production for clinical use. Many of them are semisynthetic. They are made from compounds derived from natural antibiotics and often represent clinical improvements over the natural product. Others are so similar in chemical structure and biological activity that they must be regarded as competitive drugs. As mentioned before, however, there is justification in having them available because some patients tolerate them better and some infectious bacteria are less liable to develop resistance to them.

Not all antibiotics are used principally in clinical or veterinary medicine. It has been found that the growth and development of livestock is stunted by bacteria, especially in the alimentary canal. Antibiotics that kill such bacteria improve the development of domestic farm animals and are therefore incorporated routinely into the animal feed. However, in every bacterial population, there is a small percentage of organisms resistant to this or that antibiotic. The antibiotic in the feed kills the bacteria that are affected by this chemical but leaves the resistant ones to multiply. If such resistant microorganisms get into the food chain, they can infect the human consumer of meat and then cannot be treated easily with existing antibiotics. Nevertheless, the livestock industry is loath to give up the advantages of antibiotic-enriched feed and uses one-half of all the antibiotics produced for its purposes. The greater danger lies in resistance to the so-called broad-spectrum antibiotics, which are the favorite tools of the clinical chemotherapist because they free the scientist from the task of determining what infectious agent he or she is dealing with.

One of the earliest and most unusual antibiotics was chloramphenicol, originally obtained from the fermentation of *Streptomyces venezuelae*. Its chemistry was described by Q. R. Bartz and his colleagues in 1948, and this was soon followed by a total synthesis by Mildred C. Rebstock. The synthesis was relatively simple, and now chloramphenicol is manufactured synthetically and no longer by fermentation.

Chloramphenicol is a broad-spectrum antibiotic with activity against gram-positive and gram-negative bacteria and rickettsiae. It also acts on large viruses, but its greatest value is as a curative drug in typhus and typhoid fever. One of its many side effects is bone marrow depression, which can lead to such blood disorders as aplastic anemia. Careful hematological control of patients can sidestep this danger.

In contrast to the penicillins and cephalosporins, which prevent the biosynthesis of bacterial cell walls, chloramphenicol inhibits bacterial protein synthesis from the amino acids. This activity kills bacteria but may also be the cause of the side effects of chloramphenicol on the cells of their mammalian hosts.

Macrolide Antibiotics

The macrolide antibiotics are produced by fermenting streptomycete strains that have been found in soil samples collected in all parts of the world. Those macrolide antibiotics used in chemotherapy and in agriculture are erythromycin, leukomycin, oleandomycin, tylosin, and spiramycin. Chemically they are distinguished by large rings of carbon and oxygen atoms (lactone rings) attached to unusual sugars (deoxy sugars). Their biological activity extends mainly against gram-positive bacteria and mycoplasma; they inhibit the biosynthesis of bacterial proteins at the stage of assembling amino acids to form the long peptide chains.

These antibiotics are being manufactured by competing pharmaceutical firms, whose drug insert pamphlets emphasize apparent advantages of their respective compounds. Physicians are influenced by these write-ups and often prefer the first antibiotic of a group over the latecomers. Erythromycin and oleandomycin have similar antibacterial activity spectra, and a choice between them and others that act more or less on the same infections is one of personal preference. Another factor is drug allergies of patients, who sometimes tolerate one of a group of related drugs better than another.

Tylosin is promoted especially for the control of gram-positive bacteria and mycoplasma infections of farm animals, such as swine enteritis and respiratory diseases in poultry, dogs, and cats.

Aminoglycoside Antibiotics

This group contains nitrogen-containing (aminocarbohydrate) sugar structures, several of them linked together. They inhibit gram-positive and gram-negative bacteria, including some hard-to-treat infections (such as *Pseudomonas aeruginosa*) and tubercle bacilli. Among these antibiotics are some of the earliest ones, streptomycin from *Streptomyces griseus* and dihydrostreptomycin. Streptomycin was the first effective agent for use against tuberculosis. However, its toxicity for the eighth cranial nerve causes deafness, and its use has been abandoned in the United States. The neomycins act similarly and are generally restricted to topical infections where they are not absorbed. Paromomycin, which comes from *S. rimosus*, is also effective in amebic dysentery, shigellosis, and salmonellosis. The gentamycins, from *Micromonospora*, are good for *Pseudomonas* and *Proteus vulgaris* infections. The kanamycins, isolated from *S. kanamyceticus* in Japan, are especially suitable to combat urinary infections that resist other antibacterial agents.

Tetracyclines

Chlorotetracycline was obtained from *Streptomyces aureofaciens* in 1947, and 5-hydroxytetracycline from *S. rimosus*. Soon thereafter, tetracycline itself, the parent antibiotic of the others, was prepared by fermentation. These antibiotics are chemical derivatives of a four-ring system of carbon atoms called hydronaphthacene. Their biological activity is broad. They kill gram-positive bacteria, gram-negative cocci (but not gram-negative bacilli), spirochetes that cause syphilis, tick-transmitted rickettsias (such as Rocky Mountain spotted fever), and some larger viruses, including some viruses involved in viral pneumonias.

Chemical transformations of tetracycline have yielded many different derivatives, of which doxycycline should be mentioned. These semisynthetic antibiotics have provided competing manufacturers access to the lucrative tetracycline market, and some of them are indeed a little more effective or a little less toxic than others.

Griseofulvin

Griseofulvin is obtainable from the metabolic fermentation of *Penicillium griseofulvum* and the mycelia of other *Penicillia*. It is orally active and is used for the systemic treatment of fungus infections.

Lincomycin and Novobiocin

Lincomycin, which comes from *Streptomyces lincolnensis,* has a unique chemical structure that was elucidated by a team of investigators at the Upjohn Company. It is highly effective against *Staphylococcus aureus* and other gram-positive bacteria that are spread in hospitals. It shows activity against these infections in patients who are allergic to penicillin.

Novobiocin, which comes from several *Streptomyces* species, was isolated numerous times from these fermentations and given different names. Its chemical structure is complex and consists of three separate sections joined together. Novobiocin is mainly active against staphylococcal and *Proteus*-infected skin lesions, urinary tract infections, tonsillitis, and osteomyelitis.

The Rifamycins

The rifamycins are fermentation products of *Streptomyces mediterranei.* They inhibit gram-positive bacteria and mycobacteria. Their chemistry displays an elaborate large ring of carbon, nitrogen, and oxygen atoms that has taxed the ingenuity of some of the best organic chemists; the main contribution to this complicated problem was that of Piero Sensi in Milan, Italy. He also carried out extensive molecular modifications of the natural rifamycin antibiotics and arrived at a "long-shot" derivative, rifampin. This compound has become an important anti-tuberculosis chemotherapeutic agent.

Peptide Antibiotics

More than 200 peptide antibiotics have been discovered, and the chemistry of some of the smaller peptides has been cleared up. As mentioned before in connection with peptide hormones, peptides are compounds composed of amino acids. The larger peptides with molecular weights over 6,000–10,000 are the proteins. The sequence of the amino acids in peptides determines their biological functions. The shapes of the peptide molecules also affect the biological activity—whether the peptide will be a hormone or an antibiotic or just a biologically nondescript chemical. Some of the peptide antibiotics contain additional fragments, for example, the actinomycins, which are active against tumors. Other peptide antibiotics display circles of amino acids; they have cyclic structures. Examples are the bacitracins, valinomycin, the polymyxins, and gramicidin-S. Valinomycin can put a metal ion (potassium) in the center of its peptide ring and transport it across barriers, perhaps "stealing" potassium from some cells in this manner.

Most peptides cannot cross barriers of membranes easily, and therefore their antibiotic action is most useful in dermatology, where lotions or ointments are applied topically to burns, superficial ulcers, and other lesions.

Resistance to antibiotics has been building up steadily. Resistant pneumococci and other pathogenic bacteria are encountered more and more frequently. Vancomycin, a macrocyclic glycopeptide from *Streptomyces orientalis* is one of the last antibiotics effective against many, though not all, resistant bacterial strains.

The increasing emergence of resistant strains of many infectious organisms has forced the pharmaceutical industry and the medical profession to search for "second-" or "third-generation" antibiotics that might overcome some of this resistance. The advances in peptide synthesis have also opened the door to molecular modification by inserting alternate amino acid units in the structure of peptide antibiotics. Some unrelated molecular changes could lead to new peptidomimetics that could have new uses in chemotherapy.

17

Antiparasitic Drugs

MALARIA AND ANTIMALARIALS

Malaria is caused by protozoans (minute acellular or single-celled animals) of the genus *Plasmodium*, which have a life cycle partly in humans and partly in the female *Anopheles* mosquito. Four plasmodial species cause human malaria: *Plasmodium falciparum, P. vivax, P. malariae,* and *P. ovalis*. Similar parasites infect other mammals (*P. knowlesi* in monkeys), and *P. gallinaceum* occurs in fowl and reptiles. Falciparum malaria is also called malignant tertian malaria and, if untreated, is a life-threatening disease. Benign tertian malaria (with symptoms that recur every 48 hours) is caused by *P. vivax*; it leads to relapses for years after the primary infection. Quartan malaria (with symptoms that recur every 72 hours) is caused by *P. malariae*; it is less severe than vivax malaria but can persist for many years.

Malaria is characterized by attacks of very high fever that leave the patient prostrate and debilitated. The disease is endemic in hot, humid climates conducive to the development of the insect vector, the anopheles mosquito. However, people in more moderate climates need to be concerned about malaria also. It has been stated that the Roman civilization declined because the population was weakened by malaria. In Charleston, South Carolina, and New Orleans, Louisiana, malaria was prevalent until recent times. The decisive control of malaria is achieved by a systematic campaign against the mosquito vector with insecticides. The introduction of DDT (chlorophenothane) by Paul Müller in the late 1930s did more for the eradication of malaria in southern Europe and North America than any other measure. In southern Asia, South America, and tropical Africa, the topography and economics of the regions have not made these attempts at mosquito control effective, and worldwide hundreds of millions of cases of malaria are still counted.

DDT and some other highly chlorinated insecticides kill susceptible insects such as mosquitoes on contact. Inevitably, environmental vegetation on land and in lakes and streams is covered by the insecticides and is toxic to birds and fish. In turn, birds of prey and mammals eat the fish and are poisoned. Domestic animals may feed on poisoned vegetation, and when people eat these animals or drink their milk, the poison gets into their systems. Therefore, DDT, chlordane, and similar insecticides have been banned. Unfortunately, the control of disease-causing mosquitoes, fleas, flies, and other insects has lost its efficiency through this decision.

The plasmodia undergo sexual development in the gut of the female anopheles mosquito and accumulate in its salivary gland. When the mosquito takes a blood meal, it injects a plasmodial spore (*sporozoite*) into its victim. The sporozoites rapidly assemble in the liver of the victim, where they begin asexual division that is characterized by multiple segmentation (*schizogony*). No symptoms of the disease appear for 8–12 days. Plasmodial cell division is complex; we will list only some of its stages. The plasmodial cells first divide outside of blood corpuscles. The liver cells in which they develop rupture and then empty a form called *merozoites* into the bloodstream. The merozoites infect red blood corpuscles and invade them. The blood cells rupture, release merozoites, and the merozoites reinvade new red blood cells. A few merozoites next multiply by a sexual process, and for this purpose they become male and female gametocytes. When the host is bitten again by a mosquito, these male and female cells enter the stomach of the insect, become fertilized, and produce egg-type cells, which go to the salivary glands, rupture, and release the sporozoites that infect the next victim of the mosquito.

The simplest way to stop the life cycle of the plasmodia would be to kill the sporozoites that enter the human body with the bite of an infected mosquito. This has not yet been possible. An antisporozoite vaccine might be the best measure to cope with the start of the infection.

A malarial patient has no other choice but chemotherapy. The time-honored drug from the 17th century to the recent past has been quinine. Quinine is the principal alkaloid of the cinchona tree. Its chemistry is complicated, and attempts have been made repeatedly to separate the molecular features that endow the alkaloid with suppressive activity. This effort has been the guiding thought in much antimalarial research. This work experienced three climactic peaks. The first one came during World War I, when the German pharmaceutical industry tried to synthesize drugs that might replace quinine because quinine was no longer accessible from its Dutch East Indies cinchona plantations. This effort was only partially successful, but it pointed the way for future research.

The second phase of antimalarial research occurred during World War II, after the Japanese invasion of southeast Asia and the renewed loss of the sources of quinine. From that massive and sophisticated research in the United States, the former Soviet Union, and Great Britain arose some antimalarial drugs that are in use today.

The third wave of antimalarial research arrived under the sponsorship of the U.S. Army during its ill-fated adventure in Vietnam. The U.S. Army encountered plasmodia in Vietnam that were resistant to the available drugs, and new chemicals had to be invented to overcome this resistance. In spite of many successes, the problem of chemotherapy of malaria has not yet been solved completely.

Antimalarial drugs are tested in infected animals. For decades, avian plasmodial infections in chickens and turkey chicks were used, but after World War II, an organism (*Plasmodium berghei*) was discovered that could be used in mice. Since then, in the hands of Leon Rane of the University of Miami, antimalarial tests have become faster and more efficient. Follow-up studies are made in Rhesus monkeys infected with *P. knowlesi* and in human volunteers bitten by infected anopheles mosquitoes.

The mechanism by which antimalarials counteract disease is based on an enzyme that the parasites contain. This enzyme unravels the iron-containing molecule of hemoglobin, the protein of the red blood corpuscles (*erythrocytes*) of the host. The quinoline antimalarials inhibit this plasmodial enzyme and thereby protect the host's red blood cells. These antimalarials also become layered between helical turns of nucleic acid and prevent the uncoiling needed for initiation of cell division of plasmodial parasites.

The researchers in the United States and continental Europe took their lead from one of the structural sections of the quinine molecule, a ring system called quinoline. To this quinoline was attached a nitro-gen–carbon–nitrogen chain of atoms, and the resulting compounds showed potent antimalarial activity both in bird malarias and in mammalian malarias. Of the 30,000 candidate compounds, chloro-quine has remained one of the most effective suppressive antimalari-als. It can be taken orally to ward off the symptoms of infections by the forms of the plasmodia that reproduce asexually. Primaquine is an effective drug against the sexual forms of the parasite. Quinacrine (Atabrine) was used widely by the military to keep troops free of relapses; it has the disadvantage of dyeing the skin yellow. During the U.S. Army's research effort in Vietnam, a quinine analogue called mefloquine was developed by R. E. Lutz at the University of Virginia. It is singularly active against falciparum plasmodia that have become resistant to chloroquine, but resistance to mefloquine has also begun to be encountered.

In Great Britain, the enzyme dihydrofolate reductase was found to be needed by the plasmodia for building nucleic acids in their cells, and therefore pyrimidine analogues were constructed that might have a chance to counteract this enzyme because pyrimidines are among the products made under the influence of dihydrofolate reductase. The two pyrimidine analogues, selected from huge numbers of ana-logues, were chlorguanide and especially pyrimethamine (Daraprim). In 1972, investigators in China isolated an unorthodox drug from a traditional Chinese medicinal herb, *Artemisia annua* L., which had been known for almost 2000 years as quinghao. The drug, quinghaosu (or artemisin), is apparently an effective antimalarial.

DRUGS FOR AMEBIASIS

Amebiasis is caused by infection with a parasitic, microscopic, one-celled protozoan called *Entamoeba histolytica*. This parasite has five stages in its life cycle, all of which occur in the human intestine. The infection is transferred by amoebal cysts (capsules formed around amoeba going into a resting or spore stage) in contaminated food or water, in cyst-containing clothing, or directly from an infected person. The cyst passes into the small intestine, it ruptures, and the contents disperse as a stage called *metacysts*. Each of these metacysts form *tro-phozoite* organisms, which grow, invade tissues, and form lesions by digesting the tissues of the host. Further down in the intestine, the trophozoites mature to cysts through intermediary stages and become ready to infect the next victim.

Amebiasis is characterized by dysentery and liver abscesses. It used to be regarded as a tropical disease but has also been found in more moderate climates. It is 14th among diseases as a cause of illness and 13th as a cause of death worldwide. Although the incidence of amebiasis is highest in the tropics, at least 5% of the U.S. population is infected at any given time. The most apparent causes of the spread of amebiasis are poor sanitation, lack of pure drinking water, and lack of sewage facilities, rather than geographic latitude.

Various laboratory animals can be infected with amoebas, and drugs can be tested against these infections. In animal models and in the clinical trials of antiamebic agents alike, the major difficulties are in dealing with intestinal versus liver infections.

The oldest amebicidal compounds are the ipecac alkaloids. The principal alkaloid is emetine. As its name indicates, it causes vomiting, a property for which powdered ipecac root has been used for centuries. This toxic action is minimized by concurrent administration of a sedative, and then emetine can unfold its antiamebic action. Because emetine shows a number of other bothersome side effects, molecular modification by medicinal chemists has tried to point up the antiamebic action. Only one of the hundreds of variations tried has some clinical advantages over emetine; this drug is called 2-dehydroemetine.

Other natural plant products used successfully as amebicides in humans are glaucarubin, which comes from the bark of *Simaruba glauca*, and several antibiotics, especially paromomycin (which comes from a *Streptomyces* strain). Paromomycin inhibits the incorporation of amino acids into parasital proteins.

Purely synthetic compounds have been screened as antiamebic agents in vast numbers, and a few have been introduced as clinical drugs for amebiasis. Among them are clamoxyquin pamoate, which is not only an effective amebicide but also subdues *Shigella* infections and giardiasis, an infection caused by *Giardia lamblia* that is acquired from drinking water in eastern Europe. More or less equivalent is clioquinol. This drug has been advertised as a prophylactic agent against traveler's diarrhea under the name of Entero-Vioform, but its effectiveness for such purposes is in question. The antimalarial drug amodiaquine has shown good prophylactic activity against amebiasis. It has been overshadowed by a similar chemical agent, bialamicol, which is specific for amebiasis and its complications.

A unique and very reliable antiamebic drug is metronidazole. It was created as an antitrichomonal agent, and its amebicidal potency was discovered during additional screening against other protozoa.

Arsenical drugs, investigated by Paul Ehrlich before 1910, have lost territory to less toxic agents, but one of them, carbarsone, is still used in the chemotherapy of amebiasis.

OTHER PROTOZOAN DISEASES

Protozoa other than malarial plasmodia and *Entamoeba histolytica* are also causes of major diseases of humans and domestic animals. The two most serious of these diseases are trypanosomiasis and leishmaniasis.

Trypanosomiasis is an infection spread over the tropical zones of the globe. In Africa, *Trypanosoma gambiense* and *T. rhodesiense* cause sleeping sickness in humans. These microorganisms are elongated, rapidly moving cells. They spend part of their life cycle in the tsetse fly (*Glossina*), which transmits them to humans and other mammals through its bite. Damage caused to livestock by trypanosomiasis has resulted in denying large land areas in Africa to all domestic animals except poultry. Antelopes and other wild game serve as reservoirs of trypanosomes. Again, as in the case of mosquito control for malaria and yellow fever, broad-spectrum contact insecticides offer the best prevention of trypanosomiasis. In South and Central America, *Trypanosoma cruzi*, transmitted by another tsetse fly, causes Chagas' disease in humans, mostly in children. In animals the disease is called nagana; it is caused by *T. brucei*. Other forms of animal trypanosomiasis in South America are mal de caderas, which is transmitted to horses, mules, and dogs by vampire bats. All the trypanosomes can be maintained in artificial infections in laboratory rodents and are thereby amenable to chemotherapeutic evaluation.

Early chemotherapeutic studies of trypanosomiasis relied on various azo dyes in Ehrlich's laboratory. Because these dyes colored the host tissues, noncoloring analogues were soon sought. In carefully executed studies of molecular modification, those sections of the dyestuff molecules that convey dyestuff character (*chromophore groups*) were exchanged for noncoloring portions. The outcome was suramin, which is still a valuable drug for African sleeping sickness.

Another product of the period before 1920, tryparsamide, is still used in Gambian sleeping sickness. An arsenical, melarsoprol, is more valuable as a trypanocide.

Suramin also effectively controls the nematode worm infestation by *Onchocerca volvulus*; recently it has been claimed to halt the growth of human prostate cancer.

The parasites causing leishmaniasis are *Leishmania*. These organisms are stored in dogs and gerbils and are transferred to humans by the bite of sand flies. The parasites can be established in hamsters for chemotherapeutic studies.

The chemotherapy of leishmaniasis leaves much to be desired. An antimalarial, cycloguanil, was found to have activity against South American leishmaniases.

The most aggravating protozoan disease of women in the temperate zones is trichomoniasis, a venereal disease caused by direct sexual contact. Its causative microbial organism is *Trichomonas vaginalis*. A similar parasite, *T. foetus*, invades the genital tract of cows and leads to sterility and abortion in these farm animals.

The leading drug for trichomoniasis is metronidazole, the end product of long searches in the United States and France. After the discovery of the antitrichomonal activity of metronidazole, the drug was screened against every protozoan and other parasite. Among the conditions in which it is of value is giardiasis, a protozoan infection caused by drinking water contaminated with *Giardia intestinalis*. Metronidazole is also amebicidal.

ANTHELMINTICS

One third of the human race harbors parasitic worms (*helminths*) at any time. One usually thinks of worm infections as tropical diseases, but more than 40 million Americans are also victims of parasitic worms. These worms are also a threat to all domestic and farm animals and cause serious problems to animal breeders. The most widespread human worm infestations (*helminthiases*) are schistosomiasis, ascariasis, and hookworm diseases. The most troublesome helminthiases of animals are fluke and roundworm infections.

Most people are aware of tapeworms (*cestodes*), which inhabit the intestinal tract of the host and rob the host of nutrients. They are ingested with infected diets. Fish tapeworm (*Diphyllobothrium latum*) develops in animals and people that consume raw fish. Beef tapeworm (*Taenia saginata*) and pork tapeworm (*T. solium*) are acquired from meat in the form of worm larvae. Other tapeworms of humans are the dwarf tapeworm (*Hymenolepis nana*) and *Echinococcus granulosus*.

Flukes (*trematodes*) are primitive worms. The larvae of the lung fluke (*Paragonimus westermani*) enter the body in a diet of infested crabs. The liver fluke (*Clonorchis sinensis*) and other similar flukes are acquired from infected fish, and the intestinal fluke (*Fasciolopsis buski*), from edible water plants that harbor encysted larvae of this animal.

Three species of schistosomes exist as male and female worms that invade the blood vessels of the peritoneum and other mesenteric membranes, with symptoms arising in the intestines, liver, and spleen. Other forms of schistosomiasis are caused by *Schistosoma haematobium*, which lodges in the bladder, the anus, the kidneys, and the female genitals. Humans acquire schistosomes directly from free-living cercarial forms (larval trematodes produced in a molluscan host) of the worms in water. The cercariae invade the skin and hence many internal organs, where they undergo a sexual life cycle. The eggs of the female worms are excreted with stool and urine, hatch in water to become organisms that invade snails, and there change again to cercariae, ready to infect other human hosts. Many attempts have been made to kill the snail vectors with molluscicides, both inorganic and organic chemicals. However, extensive irrigation canals and the habit of standing in water while planting rice in many tropical and semitropical countries have made snail control difficult. Therefore, chemotherapy of established infections is of great importance. *Schistosoma mansoni* can be grown in mice and *S. japonicum* in monkeys for the evaluation of antischistosomal drugs.

Nematodes (*roundworms*) look more like real "worms." They include hookworms, whipworms, and pinworms. They invade the human skin or are acquired in the diet; *Trichinella spiralis*, which causes trichinosis, is ingested with poorly cooked, infected pork.

The variety of parasitic worms requires a variety of chemotherapeutic drugs to combat the often terrible diseases caused by the worms. Some small worms are so translucent that their internal organs can be studied from the outside. Others are covered with tough membranes, and their organs are differentiated extensively. The fact that the vectors of the worms are associated with careless and unsanitary life habits and with poverty contributes to the difficulty of chemotherapeutic intervention. In addition, no persuasion or education can be applied to domestic and farm animals that get worm diseases in the normal course of their lives. Cats, dogs, cattle, sheep, and horses need frequent deworming or anthelmintic chemotherapy for economic as well as sanitary reasons. A large effort has therefore gone into the study of anthelmintic drugs. The enormous evolutionary difference between the parasitic worm and the higher animal host points to the possibility of a comfortable margin of toxicity of drugs for the two types of organisms. However, worms and mammals depend on the same fundamental sources of energy and on the same or at least similar enzymes for the basic reactions of their life processes. If a drug inhibits a controlling enzyme in a worm, it may also affect the analogous enzyme system in a host. Therefore, drugs for helminthiases share the potential toxicity with many drugs for other types of parasitic invasions.

This connection can be seen in drugs for tapeworm diseases. These diseases are caused by several species of *Hymenolepis* and *Taenia*. One of the oldest drugs, still in use despite its toxicity, is extract of male fern (*Dryopteris filix-mas*), which contains such taenicidal compounds as filixic acid. Several other tapeworm-removing medicines have been discovered by screening in small animals such as mice and larger animals such as cats and dogs. Among them are the old antimalarial quinacrine, the phenolic derivative dichlorophen, the salicylic acid derivative niclosamide, and the antibiotic paromomycin.

In the trematode worm infection schistosomiasis, one has to combat various human blood flukes of the genus *Schistosoma*. Originally treated with the objectionable antimonial agent tartar emetic (also known as antimony potassium tartrate), schistosomiasis is still treated with antimony compounds such as stibophen and stibocaptate, but these injectable drugs have been replaced by orally active, metal-free schistosomicides that are less toxic. Two of these schistosomicides are lucanthone and hycanthone; hycanthone is a mammalian metabolite of lucanthone, and administration of lucanthone is therefore, after delay for oxidative metabolism in the body, more or less equivalent to taking hycanthone directly. Another drug, niridazole, is effective against several but not all species of *Schistosoma*. This specificity also limits the effectiveness of the other antischistosomal agents, and careful morphological diagnosis of the parasite should determine the drug for prescription in chemotherapy. Because more than 300,000 people are afflicted by schistosomiasis worldwide, early diagnosis is a serious problem.

These numbers assume even more frightening proportions in the case of the estimated 200 million people suffering from filariasis and the 30 million people who have gone blind from onchocerciasis. Other nematode infections are found in equally high percentages of the rural population in warm climates: 20% are suffering from ascariasis, 25% from ancylostomiasis, and 10% from oxyuriasis. In spite of major research efforts in a few institutions and industrial laboratories, chemotherapeutic work on nematode infections has not been one of the major competitive programs because the poverty-stricken patients cannot be expected to return enormous research investments. It is different in the husbandry fields, where nematodes wreak damage. When properly treated, the cured animals compensate for the cost of the necessary research expenditures.

Filariasis occurs in Africa, China, India, Japan, the East and West Indies, and Central America. The most important nematodes causing the infections are *Wuchereria bancrofti, W. malayi,* and *Onchocerca volvulus.* These worms go through a life cycle of which one phase is called microfilariae. These microfilariae develop into infecting larvae in mosquitoes and are transmitted by the insect to the mammalian host. There they migrate into the lymphatic vessels and continue a cycle of sexual reproduction, resulting in adult worms. If repeated heavy infection occurs, the larvae and adult worms can block lymphatic vessels and give rise to such swellings as elephantiasis. The preferred chemotherapeutic agent for the removal of microfilariae is diethylcarbamazine, which was discovered by S. Kushner and his colleagues. The destruction of the microfilariae especially in loiasis (*Loa loa* infection) may cause allergic reactions. These allergic reactions can be controlled by giving smaller doses of the drug initially or by coadministering an antihistaminic agent. Diethylcarbamazine is not very effective against onchocerciasis (*O. volvulus* infections). Onchocerciasis is controlled better by suramin.

Most patients who have intestinal-worm infections suffer from multiple infections, often infestations by *Ascaris lumbricoides* and *Trichostrongylus orientalis.* In such cases, thiabendazole is of value.

Thiabendazole has become primarily an agent for the control of gastrointestinal nematode infections of ruminant animals.

Ascariasis and oxyuriasis are treated with a simple chemical called piperazine. Piperazine is highly active, virtually nontoxic, and cheap. It was discovered by screening after trials in unrelated rheumatic disorders. Oxyuriasis is caused by *Oxyuris* (pinworms); it is the most common worm disease of American schoolchildren. Piperazine is the most effective drug for this disease. Coadministration of another anthelmintic called pyrantel, which was developed in the Pfizer Research Laboratories, clears up concurrent ascaris infections. Thiabendazole is also of value.

The Belgian drug levamisole is a broad-spectrum anthelmintic active against a large number of pathogenic nematodes in 13 hosts, including humans. It is not effective in intestinal *Trichuris* infections.

Anthelmintic drugs act by depriving the various life stages of the worms of essential enzyme systems. Manipulation of low dosages and coadministration of low (barely effective) doses of two or more drugs with overlapping activities offer the best hope for minimizing systemic toxicity, with its inherent unwanted side effects.

Many anthelmintics with some activity against exotic worm diseases have been omitted from this discussion because few physicians in the developed countries would encounter these infections and learn how to treat them. Those doctors who do will probably try out the existing newer anthelmintics or reach back to some older ones if they are still available, and they may have to undertake a careful differential diagnosis if their patients have acquired the worm infestation during a trip to unsanitary regions. Then it will be necessary to search the literature for records of clinically successful drugs for the disease at hand.

ANTIFUNGAL AGENTS

For the purposes of the chemotherapist, the term *fungi* covers both yeasts and molds that attach themselves to tissues and draw their nutrients from the cells of their hosts. The diseases are called *mycoses*. There are three types of mycotic diseases: superficial contagious skin infections (*dermatophytoses*); candidiasis, also called moniliasis, which causes lesions on the skin and mucous membranes, but can also invade internal organs; and deep mycoses that are not contagious but invade the skin, lungs, lymph, and other internal systems. Mycoses can be sand-borne, or they can arise from contact with fungi on vegetation or other objects, or from diets contaminated with saprophytic molds or mushrooms (those that obtain nourishment osmotically from the products of breakdown and decay).

A fourth class, not strictly fungus infections, is actinomycoses. They infect the lungs and other organs. They are caused by Actinomycetales, an order of organisms between molds and bacteria.

Fungal diseases were known before the early bacteriologists Louis Pasteur (1822–1895) and Robert Koch (1843–1910) recognized bacteria as causes of many infections. Fungal diseases are assuming greater importance today, not only because more cases of serious primary mycoses are being diagnosed, but because systemic fungal infections so often lodge in patients with *neoplastic* (tumorous) diseases and lead to the death of the individual. Likewise, AIDS patients frequently become victims of fungal diseases after the human immunodeficiency virus has abolished immunological resistance in patients.

Superficial fungi use keratin as their substrate; however, the fundamental biochemical reactions—the oxidative transformations yielding energy for all cells, with the same overall enzymes as in other cells—are also used by fungal life processes. Consequently, drugs for fungistasis must be found by trial and error among chemicals that inhibit cellular life in general.

Salts of fatty acids have long been claimed to be fungistatic, but their value is not beyond dispute. The best is the sodium salt of undecylenic acid, an 11-carbon unsaturated acid. It is effective in bringing to a halt the spread of tinea pedis (athlete's foot), but in other dermatomycoses on the head and in the hair it is of less value.

Other synthetic substances with a broader antifungal spectrum include miconazole, a rather nontoxic drug whose chemistry is similar to that of antihistaminic agents. Indeed, several antihistaminics have minor antifungal activities. A number of superficial fungal diseases that affect the skin and mucous membranes such as the vagina respond well to miconazole. Other topical fungal infections such as those caused by *Trichophyton* species are treatable with tolnaftate. Compounds that withdraw metal ions from essential fungal life processes can stop the reproduction of some fungi; clioquinol is one of these products. It is still used in *Candida* and *Monilia* infections but is giving way to more specific antibiotics.

Most of these antibiotics are prescription drugs because their indiscriminate application in insufficient amounts to inflamed membranes might lead to drug-resistant fungi. People suffering from minor fungal skin infections often want to relieve the itching and other discomfort without consulting an expensive physician. They turn to older, less effective over-the-counter salves and ointments, which may or may not work on their undiagnosed skin lesions. Among these agents are salicylic acid ointment, acrisorcin cream (which may produce hives and blisters), chlordantoin (for candidiasis), and several old dyestuffs that have survived with an undeserved reputation for antifungal activity.

The modern dermatologist prefers to try antibiotics for the treatment of both superficial and deep-seated mycotic infections. Griseofulvin, a product of *Penicillium griseofulvum*, is active orally and diffuses to the site of skin and hair lesions. It has low toxicity. Griseofulvin is bound to the lipids of the fungal cells. It is effective in fungal diseases of the skin, hair, and nails caused by *Microsporum, Trichophyton,* and *Epidermophyton* species. Nystatin is an antibiotic made from *Streptomyces noursei*; amphotericin B is another one made from *S. nodosus*. These two medicines belong chemically to the class of polyene antibiotics, which features large rings of carbon atoms interrupted by some oxygen atoms. They bind to the sterol lipids of yeasts and fungi, disrupt their cell membranes, and allow small nutrient molecules to leak out. Nystatin is less toxic than amphotericin B; they are both effective against a wide variety of pathogenic fungi and yeasts. Nystatin is the preferred drug for *Candida* infections, whereas amphotericin B can overcome internal mycoses that were fatal before the introduction of this antifungal.

18

Antiviral Drugs

Viruses include a vast variety of agents that manifest their "life" cycles at the expense of the cells of their hosts. They are at the borderline of animate and inanimate matter. They consist usually of a core of nucleic acid material covered with a protein capsule, but their composition can be more complicated. Both the nucleic acids and the protein coat are characteristic of a given virus.

Viruses cause untold diseases in every kind of living organism. Bacteria fall prey to viruses called *bacteriophages*; other viruses infect insects and plants. In fact, the first incisive chemical study of a virus looked at the tobacco mosaic disease virus; it was carried out by Wendell Stanley (1904–1971) in 1936. Within the framework of this volume, we are most interested in viruses that attack mammals, especially humans. There are more of these viruses than one would expect. Many people have consulted their physicians about some strange discomfort and have been told, "Its cause must be a virus." In many cases the doctor was right. Every pharmaceutical manufacturer would give his or her birthright for a drug that would cure the common cold, but it is unlikely that *one* such drug will be found because viruses have an uncanny ability to mutate and to become resistant to a given chemical. When we get our annual flu shots, we are made aware of the fact that this vaccine subdued last year's influenza virus, whereas this year's organism has not yet been identified.

However, there have been some great triumphs in viral immunization. Smallpox, caused by the variola virus, has disappeared from the globe. The dreaded annual outbreak of poliomyelitis can now be fully controlled by one simple oral immunization that is available to everybody. Furthermore, the poliomyelitis virus has been purified and beautifully crystallized. What is this virus, an infectious organism or a chemical? In fact, it is both.

The manufacture of vaccines provides one of the oldest examples of biotechnology. Blood samples from individuals infected with a given virus, or from animals artificially infected with the same virus, are collected, and the serum is incubated to promote viral multiplication. The poliomyelitis virus can be grown in monkey kidneys, and other viruses thrive in the *chorioalantoic* membrane, which separates the chicken egg yolk from the living protein of the egg. The virus is concentrated from these growth media and attenuated, that is, weakened by heat or chemical means. It is then injected into the person or animal to be protected. After a while, the body's immune defenses respond to the intruder and simulate having fought off a natural virus infection. The resulting immunity can be temporary as in influenza immunization. Among the earliest viral vaccines were those against smallpox (developed by Edward Jenner (1749–1823)), rabies (developed by Pasteur), and yellow fever (developed by Walter Reed (1851–1902)). Vaccines are not restricted to viral diseases; for example, pneumonia caused by a pneumococcus can be averted by long-term immunization.

When a virus approaches the cell of a host that is to be its victim, it sheds its protein coat. Its naked nucleic acid core attaches itself to the host's cell wall. The nucleic acid core penetrates the cell wall and opens housekeeping inside the cell. The virus starts to digest (one of the definitions of life) the cell's chemicals and directs the cell nucleus to obey its demands by producing more viral nucleic acids instead of the nucleic acids of the host. The viral nucleic acids thus manufactured by the intruding dictator escape and attack new host cells.

If we want to prevent this sequence of biochemical events, we can try to interfere with one or all of these steps. We can even go further now. The virus needs certain enzymes for these steps as well as enzymes to synthesize its protein coat, and we can try to block these enzymes. Some large viral proteins must be broken down to smaller proteins with a regulatory function by enzymes called *proteases*. Interruption of this step could check the disease. Several experimental drugs that inhibit viral proteases are being studied. However, we have to be careful not to inhibit the host's enzymes as well, which are needed to synthesize normal chemicals of the cell membrane and the cell interior; they are often not very different from those used by the virus. This similarity can represent toxicity to the host.

Some viruses differ from mammalian cells in that they reverse the usual processes of nuclear replication. There are two types of nucleic acids: deoxyribonucleic acid (DNA) and ribonucleic acid (RNA), named according to the carbohydrate component (deoxyribose or ribose) contained in their molecules. Cells undergo division (which is also called multiplication) by unfolding the DNA helix and having their RNA copy the arrangement of the DNA composition. In a series of further steps, the RNA directs amino acids to arrange themselves as proteins characteristic of the species. That is how species specificity is maintained, how humans reproduce human children, and orangutans reproduce orangutan babies.

Some viruses do not follow the DNA→RNA sequence. They start replication by a reverse sequence, RNA→DNA. Such viruses are called retroviruses; among them are herpes viruses and the human immunodeficiency virus (HIV), which is the cause of acquired immune deficiency syndrome (AIDS). The reverse replication sequence, RNA→ DNA, requires some enzyme catalysts not present in the normal cell that replicates by the DNA→RNA scheme. These retroviral enzymes have been the target of most drugs designed to block the retro-replication reaction.

Nucleic acids are large molecules, ranging from a few dozen to many millions of building blocks called nucleotides. Drugs that are designed to interfere with reactions of these nucleotides are similar in chemical structure to portions or the whole of these nucleotides. Among them are zidovudine (AZT), didanosine (ddI), zalcitabine (ddC), and stavudine (d4T). The names in parentheses are acronyms of the chemical names of the compounds, such as 3'-azidothymidine, AZT. Some of these drugs are less toxic than AZT but no more effective. None of them work on the mysterious phase of retroviral life processes, that is, the dormant state of the virus. During the dormant period, the virus appears to retreat into nerve fibers and stay there for long periods of time before it breaks out and becomes truly virulent, suppressing the immune system and causing the fatal stage of AIDS. The process of this viral retreat into nervous tissue is not yet understood.

Hundreds of thousands of chemicals have been screened for activity against HIV viruses, and quite a few have been selected for further study, but so far with little clinical success. The rapid rate of mutation of the viruses complicates these studies. This mutation can be illustrated with another viral disease, influenza. There are mainly two types of influenza, called A and B. A drug called amantadine is somewhat active against influenza A but fails to suppress the influenza B virus in spite of the close relationship of these two viruses.

Some viruses are larger than others. Among the larger ones is the virus that causes viral pneumonia, and in such cases chemotherapeutic antibacterial antibiotics may have some effect. Tetracycline, a broad-spectrum antibacterial antibiotic, is an example of chemicals that also subdue larger viruses. It is interesting that even a plant virus, the virus that kills coconut palm trees in Florida, can be treated with injections of tetracycline into the bark of the tree.

The overlapping of normal cellular and some viral life processes finds an analogy in a similar overlapping of malignant tumor cells and normal host cells. For this reason, compounds tested in cancer chemotherapy are routine candidates for inhibition tests of viral cell processes. Because cancer chemotherapy dates back 40 years, a considerable collection of surviving chemicals has accumulated and become available for the more recent projects in antiviral chemotherapy. If one wishes to take advantage of this "library" of potential antiviral drugs, one must find some rationale in selecting the most likely candidates. One of the options is to correlate structural features of drugs by programming such comparisons in computer searches.

19

Computer Assistance in Medicinal Research

The main outward difference between humans and animals is the human's ability to speak. Writing down one's thoughts is an extension of one's efforts to preserve the spoken word. The machines called computers invented in the 20th century further extend and speed up this process. By inserting a tape or disk on which a given type of information has been imprinted by transistor techniques (*computer software*), one can program the computer to answer questions in certain fields (such as stock prices, statistics of labor markets, flight plans of aircraft, properties of drugs, and results of immunizations). However, computers are incapable of creating thought processes. A generation ago, before electronic pocket calculators became commonplace, everybody added, subtracted, multiplied, and divided skillfully on a piece of paper or in their heads. Now many children would not know how to perform these tasks without calculators. The same thing is true about computers. They speed up our activities, they enable us to get results in seconds that would have taken a lifetime earlier, and they thereby free us to allot our time to other tasks. But computers cannot replace ingenuity, keenness of observation, and intellectual experience.

In drug research, computer assistance has become important in several fields. First of all, most drugs have to be synthesized in adequate amounts for experimental testing and for clinical uses. The synthesis of structurally simple drugs is straightforward and offers few if any developmental difficulties. However, drugs with complicated structures such as rare natural products can be synthesized by several routes. Several routes call for widely different starting materials, synthetic pathways, and chemical reagents. The decision about which of these materials and pathways should be employed depends on cost estimates of reagents, yields, availability of facilities, expertise, and other factors. Computers can aid in choosing the most advantageous approach. The Nobelist E. J. Corey of Harvard University has published examples of computerized applications to organic synthesis.

In choosing molecular modifications of existing prototype agents, one has to consider many chemical and physical properties of the proposed compounds. Will the new analogues be acceptable on the basis of predictable data? Such questions can be answered readily if suitable statistical data stored in a computer are consulted. However, the hope expressed by foes of trials in laboratory animals that such trials will be replaced by computer-aided predictions can barely be given credence. The information that researchers can learn from the living organism of a laboratory animal has not yet been rivaled by computer software, and it is doubtful that it ever will.

The choice of suitable chemical substrates as agonists or antagonists of enzymes and other biological macromolecules has been and will be facilitated by computer modeling. For example, in the case of potential antiviral agents, the chemical structures of candidate compounds are projected by three-dimensional devices on a computer screen, with lines representing chemical valence bonds. Then the structures of enzymes (if known) or of characteristic viral cell chemicals (such as nucleotides) are projected on the same screen. The two pictures, that of the candidate drug and that of the viral chemical, are superimposed where possible. If any significant overlapping of the two line drawings can be achieved, the drug may be selected for screening.

Such computer-aided searches for candidate chemotherapeutics greatly reduce the time and effort that would be required otherwise. Because screening even for preliminary chemotherapeutic activity is expensive, the computerized search is much more economical than laboratory trials. This method has now been applied for candidate drug selection in most fields of drug design. The software for this work contains new programming language to simulate mechanisms by which a given type of drug might act.

Pharmacologists who test chemicals in mice, larger animals, and often in expensive primates will profit from computer-stored knowledge of the metabolism of each animal species and data concerning previous experiments with drugs similar to the proposed agent to be tried. Effects of the drug under study on isolated organs and on specific enzymes in vitro can be factored in when planning new methods of animal experimentation. In all these activities, time will be saved by using computers, biological safety will be enhanced, and the rationale for a given trial will be strengthened.

All chemical and biological studies are preceded by literature searches. The purpose of this library work is to find out whether any similar experiment or procedure has been recorded previously, how it may differ from experiments now being planned, and how pitfalls may be avoided. The literature of biology and chemistry has grown so much in the last six decades that such searches, even if based on existing excellent general indexes, can be extremely time-consuming and physically exhausting. Computers can cut these efforts greatly, especially if journal articles, monographs, and reviews have been stored in the instruments. Students now use computers routinely in solving problems and scanning the contents of textbooks and other sources of pertinent information. Medicinal scientists emerging from universities are now inevitably prepared to apply computers and suitable software to the problems confronting them in their professions.

20

Antiseptics and Disinfectants

The air we breathe and the water we bathe in are loaded with "germs," that is, bacteria, fungi, and other microbes. The walls and floors of our buildings, the clothes we wear, and the dishes on which we serve our food are equally covered with microorganisms. So are our bodies, even if we wash them thoroughly. That is why surgeons scrub and cover their hands with surgical gloves before they operate. Some of these microorganisms are harmless, and some are *pathogenic* or destructive because they feed on tissues and materials on which they alight. If we wish to live hygienically and protect our dwellings and surroundings from pathogens, we must cleanse ourselves and the objects around us with chemicals that inhibit or kill the damaging microbes. Some nonpathogenic bacteria and fungi decompose perspiration and fibers to find nutrients; they produce offending odors and discolorations. A large industry has been built up to manufacture chemicals, soaps, and preparations to combat microbial processes on inanimate surfaces and on mammalian skins. The advertisements for such chemicals point out unhesitatingly the damaging consequences of *not* using those products. Commercial television and newspapers depend on the advertisements for their economic survival.

Sterilization means destruction of all forms of life, whether vegetative cells of bacteria, bacterial or other microbial spores, or viruses. Long exposure of surgical instruments to sterilizing solutions at high temperatures may achieve sterilization, but larger objects cannot be fully cleansed of microbial cells. Such objects as floors, walls, and toilets can only be disinfected. Disinfectants kill most vegetative forms of microorganisms but not bacterial or fungal spores, and they cannot remove viruses from inanimate objects. Sanitizers must reduce bacteria to levels acceptable to public health rules. They are of significance in cleansing plates and cups in restaurants and in cleaning machinery in food-processing plants. If chemicals only inhibit the multiplication of microbes, they are microbistatic (also bacteriostatic, virustatic, etc.). If they apparently kill the microbes, they are called microbicidal (bactericidal, virucidal, fungicidal, etc.). Such -cidal agents, when applied to living tissue, are termed antiseptics, such as mouthwashes and deodorant soaps.

In commercial disinfectants and antiseptics, the active ingredients are mixed with detergents, antifoam agents, dispersants, aerosols, and coloring matter. Occasionally, more than one active ingredient is present. The antibacterial potency is determined by the killing power of the agent for a number of representative microorganisms and is compared to an arbitrarily chosen standard. For example, a certain concentration of phenol (carbolic acid) is given a value of 1.0, and the dilution of the disinfectant to be tested that kills test bacteria in a certain number of minutes is assigned a comparative value. The bacteria that must be killed under these conditions are *Salmonella typhosa* or *S. cholerae suis*, *Staphylococcus aureus*, and *Proteus vulgaris*. The standard fungus is *Trichophyton interdigitale*. Other tests measure comparative activity against some bacterial spores, tubercle bacilli, and specific viruses against which effectiveness is claimed.

The following types of disinfectants are the most common: phenolic compounds that are acidic and go into solution in alkali; quaternary ammonium compounds that are water-soluble ionic materials; oxidizing agents, such as hydrogen peroxide; volatile alkylating agents, such as ethylene oxide, formaldehyde, and propiolactone, which can sterilize inanimate objects by fumigation; chlorine, chloramines, and hypochlorites, which are used in swimming pools and in household sanitizers; and miscellaneous substances that can be used dermatologically and border on topical chemotherapeutic agents.

Phenol (carbolic acid), used in 1867 by Joseph Lister (1827–1912), is only of historical interest because of its toxic, necrotizing properties (properties that cause localized death of living tissue). A number of related compounds, obtained from coal tar or by complete synthesis, have been more useful. They contain chlorine or various carbon chains (alkyl groups). Among those compounds in use in antisepsis are chlorocresol, chloroxylenol, metacresol acetate, hexylresorcinol, and hexachlorophene. Some of these products are used as skin disinfectants, in the outer ear, in mouthwashes, and in dermatology. Oxine is a phenol derived from a nitrogen-containing system called quinoline. It can bind metal ions, and by removing them, it deprives microbes of ions that are essential to their life processes.

If an atom of nitrogen is combined with four organic chemical radicals, and if one of these radicals is long enough to make the compound look like a soap, quaternary ammonium salts ("quats"), which have potent antiseptic activity are the result. Quats are not without drawbacks; the most serious drawback is that some bacteria become resistant to them. Some quats are also effective antiviral disinfectants. Among these quats are cetylpyridinium chloride, benzethonium, and benzalkonium.

Chlorine and hypochlorites dissolve in water to give a certain amount of hypochlorous acid, which kills microbes by oxidation. Because other organic matter (such as plant debris) is also oxidized, it is necessary to clean the water before applying a source of hypochlorous acid. A number of organic chloramines such as halazone, chloro succinimide, and chlorinated isocyanuric acids are more versatile agents for the release of chlorine. They are effective wound disinfectants, but may be toxic when ingested.

Chlorine is not the only bactericidal halogen. Iodine tincture acts similarly in solution, owing its activity to the combination of iodine with bacterial proteins. A number of peroxides, perborates, permanganate salts, and other oxidizing agents are used as disinfectants on various body surfaces. They oxidize (burn up) components of bacterial membranes and other proteins.

Overall, antiseptics and disinfectants are used to make our world cleaner and safer. We maintain a vigil against the forces of dirt and "germs" with our tools for sanitation and hygiene.

21

What's Next?

Recent trends of research on new drugs have been dominated by efforts to simplify drug design by basing it on biochemical interactions of chemicals with drug receptors. Because most drug receptors are complex proteins, biochemical experts on proteins lead this field of inquiry. Progress in receptor study has resulted from experiments that at first glance have nothing to do with biomedical problems. This experimentation emphasizes the freedom needed for unconventional research projects; they may not seem to have practical relevance, but their application to medical science has often proved successful. Unconventional research projects have resulted in advances in instruments that measure shapes, sizes, electronic behavior, and other properties of large molecules. Thus, experts on mass spectrometry (which measures the total weight of a molecule), experts on computer-aided X-ray diffraction (which furnishes the length, width, and height of a molecule and of its parts), and electron microscopists and nuclear magnetic resonance (NMR) spectroscopists have made important con-tributions to the understanding of drug receptors.

Once the structure of the most reactive sites on the surface of the large receptor molecules is known, one can attempt to overlay these reactive sites in computer graphics with chemicals that could activate or block those sites. In many cases these chemicals have been chosen from among proteins. Dozens of small companies have sprung up that specialize in such interactions. The difficulties of their studies have led to many failures, and only few of these new companies have survived. In some cases, large established pharmaceutical industries have found it more convenient to buy up small and financially struggling enterprises than to assemble their own research divisions in molecular biology.

The expertise of molecular biologists has also been applied to other areas, such as molecular engineering and biotechnology. The first biotechnology drug was recombinant human insulin. This insulin was produced by changing the genetic underpinnings of microbes such as the colon bacterium to metabolize its amino acids to secrete human insulin in a process adaptable to mass production of this anti-diabetic hormone. Other bioengineered drugs already in use are the human growth hormone for the treatment of dwarfism in children; the intrinsic factor, which is needed to avoid pernicious anemia; and difficult-to-obtain anticancer drugs.

The primary call for new drugs comes from the medical profession and from the public. Hardly any of the hundreds of current drugs are fully satisfactory, and all of them could stand improvement. Side effects often limit the use of drugs to around 80% of the patients, and these 80% are not immune to minor but bothersome drawbacks. Unless drugs with virtually complete specificity are discovered, the demand for medications without side effects will have to be put on hold for the foreseeable future. Even drugs like the penicillins, first believed to be totally specific as antibacterial agents, are not tolerated by one-fifth of the world's population because of an inherited allergic response to these agents.

Resistance to all kinds of drugs, not only to antibiotics, calls for alternate agents that can confront diseased cells with new antagonistic activities to which the cells have not yet become refractory. The public periodically cries out for a drug therapy for its most feared diseases. Cancer, AIDS, and genetic defects are mentioned most often. The public is not aware of ongoing research in these fields, and they appreciate only to a small extent the difficulties confronting scientists in these areas. They do not know that 88 new drugs for AIDS were in development in 1994, among them the antiviral didanosine (ddI), and that seven vaccines are in clinical trial.

Predictions about future trends in drug development, especially in the pharmaceutical industry, are fraught with difficulties because of "fashions" in medicinal research. Everybody agrees that a biochemical basis (as in the understanding of receptors) is a must in developing more rational foundations for drug discovery. Receptor studies are conducted often in nonindustrial research institutions such as the National Institutes of Health, privately and governmentally supported laboratories, and academic research units. When new insight into such studies is gained, the industry usually expands this work in a specific disease-related drug development. Such events come in waves because published biochemical studies become available simultaneously to all industrial scientists. Therefore one could extrapolate from the published findings which way drug developments may go. In a few instances, fundamental biochemical discoveries have been made in the industry, and the particular corporation that made the discoveries thereby holds an advantage over its competitors.

However, published findings in one field do not translate into a change of direction in research for the whole biomedical field. For example, a detailed description of receptors for neurotransmitters produces a wave of innovative drug development in companies that specialize in drugs for neurological disorders, but other companies, which are primarily interested in drugs for infectious diseases, remain less involved in this research.

Research on new "second- and third-generation" antibiotics will proceed unabated because of the development of resistance of many pathogenic microbes to existing antibiotics. There was a period of relative slowdown in this field from 1980 to 1990, with few companies mounting research programs for new and really different antibiotics. As more and more of the older agents fail to provide protection against deadly infections, however, the industry will reenter the field in full force.

The pharmaceutical industry has to return a profit to its investors and therefore finds it more advantageous to concentrate drug research and development on diseases that affect large numbers of patients and furthermore, patients whose health insurers will pay for drug therapy. Many infections and infestations ruin the lives of hundreds of millions of sufferers who cannot afford to pay for drugs of any kind, especially in underdeveloped regions. Therefore, most research industries cannot support investigations of drugs for these "orphan diseases". Some governments of industrialized countries have provided encouragement for such researches by tax exemptions on expenses and profits for a number of years. Still, only few companies have taken advantage of these provisions. It will take a long time for native industries in developing nations to attain the necessary expertise to do their own R&D work on diseases endemic in their own countries.

Almost every news organ for scientific readerships bemoans the level of public financial support of fundamental research, especially in the field of health care. But money is not everything in catalyzing new methods, discoveries, and diagnostic advances in therapy. Billions of dollars have been appropriated for cancer research for more than 40 years, billions for viral diseases and especially AIDS for 15 years, and more billions on genetic diseases, with only partial success in cancer and not enough in viral and genetic afflictions. Money can be used to buy a very expensive instrument for a gifted investigator and to allocate funds for facilities, co-workers, and technicians, but unless one scientist has one fundamentally novel idea or stumbles onto a decisive new approach, funding alone will not give us new cures for dreaded diseases.

How can we augment the small number of such exceptional, inspired scientists? The answer lies in supporting the education of gifted boys and girls in grade school and high school by separating them from the run-of-the-mill curricula that were designed to give a measure of literacy to everyone and by guiding the gifted youngsters into the physical and life sciences and mathematics at the expense of other endeavors. Upon entering college, which should be free for such candidates, the general course work should be waived and the student should be guided rigorously into sciences and prepared for postgraduate research. This process of selection of a scientific elite will pay off without doubt in accelerating new discoveries in therapy.

Glossary

Some Chemical and Medical Terms

Alzheimer's disease   Dementia caused by deterioration of brain cells with production of plaques.

Anticonvulsant   A drug that counteracts or prevents convulsions caused by brain diseases, electric shock, and certain chemicals.

Antihistamine (antihistaminic)   A drug that counteracts damaging effects of histamine, such as allergies and excess stomach acid.

Antihypertensive   A drug that lowers elevated blood pressure.

Antimalarial   A drug useful in the chemotherapy of malarias.

Antimetabolite   A chemical of a structure related to but not identical with a metabolic biochemical. If it counteracts the effects of the metabolite, it may become a useful drug.

Antipyretic   A drug that lowers elevated body temperature.

Antispasmodic   A drug that blocks spastic effects of neurohormones, relaxing muscles and preventing excessive secretion from certain glands.

Antitussive   A drug that prevents coughing.

Arrhythmia   Irregularities in rate of heartbeat.

Arthritis   Inflammation of joints. May be a degenerative disease, perhaps related to loss of immune response.

Bacteria   One-celled microbes without chlorophyll that multiply by cell division. Spherical bacteria are called cocci; rod-shaped bacteria are called bacilli; and spiral bacteria are called spirilla. Some bacteria are pathogenic, others are beneficial, for example, in nitrogen fixation.

Bactericidal   A drug that kills bacteria.

Bacteriostatic   A drug that slows or prevents the multiplication of bacteria.

Biochemistry   The chemistry of substances, processes, and reactions in living organisms. Biochemistry can also be studied under laboratory conditions in the absence of a living organism.

Biosynthesis   Synthesis of a chemical by cells or by a laboratory process analogous to that in living tissue.

Blood pressure   Pressure exerted by the blood against the walls of blood vessels. Diastolic blood pressure is the pressure caused by the rhythmic dilation (diastole) of the heart; systolic blood pressure is the pressure during the contraction of the heart after the dilation. A normal reading would be 110 (systolic)/80 (diastolic).

Cancer   Malignant cells or tissues; cancer is at least 100 different diseases. Cells invade tissues by metastasis.

Cancer chemotherapy   Combating the spread of malignant cells with drugs.

Cardiac drug   A drug that regulates functions and contraction of the heart.

Cathartic (laxative)   A chemical that stimulates intestinal peristalsis and relieves constipation.

Chemotherapy   Drug treatment of disease-producing foreign cells such as tumor cells, bacteria, viruses, and protozoa. Sometimes chemotherapy designates any drug treatment.

Coating, enteric   Encapsulating a biologically active chemical in a layer such as of wax or biodegradable plastic that delays and regulates the rate of release of a drug into tissues.

Conditioned avoidance response   Behavioral test based on acquired learning.

Convulsion   A violent involuntary spasm of the muscles.

Cortex, cerebral   The outer layer (gray matter) of the brain.

Cough reflex   The physiological response to irritation of bronchial mucous membranes. This reflex involves complex interaction of the central and peripheral nerves.

Curare   A tarry mixture of substances obtained from the bark of certain tropical vines. It causes paralysis of the muscles and is used medicinally to reduce muscular rigidity.

Dendrite   A fine, branched protrusion of nerve cells involved in transmitting nervous impulses.

Diabetes   High blood-sugar levels (also known as hyperglycemia) resulting from lack of insulin.

Diagnosis   Process of deciding the nature of a disease by examination.

Diuretic   A drug that increases the secretion and flow of urine. Some diuretics also lower blood pressure.

DNA   Deoxyribonucleic acid, characteristic of a given organism.

Drug insert   The paper in a drug package that contains a description of the drug, its main effects, side effects, toxicity, dosage form, and names.

Drug receptor   A protein or nucleic acid that recognizes and reacts with a drug and initiates a biological effect.

Drug resistance   The ability of a cell or organ to ward off the effect of a drug.

Drugs, chiral   Drugs with unsymmetrical molecular structure. Chiral drugs exist as enantiomers.

Drugs, generic   Drugs not (or no longer) the property of a patent holder. They can be manufactured and marketed by any suitable agency.

Edema   Abnormal accumulation of fluid in cells, tissues, or body cavities resulting in swelling.

Elixir   A solution, often an alcoholic tincture, of drugs.

Enkephalins   Small-molecular protein neurohormones found in many tissues. They participate in modifying the transmission of painful stimuli.

Enteric coating   *See* Coating, enteric.

Enzyme   A protein produced by living cells that has the catalytic ability to regulate the rate of biochemical reactions. A few synthetic substances acting like enzymes have been discovered.

Epilepsy   A chronic disease of the central nervous system characterized by convulsions and unconsciousness.

Euphoria   A feeling of well-being. It can be induced by certain drugs.

Extrapyramidal symptoms   Facial rigidity, tremors, and drooling.

FDA   U.S. Food and Drug Administration.

Fibrillation, cardiac   Heart flutter and irregular beats.

Filariasis   Disease caused by nematode worms.

Free radical   Unstable atom or group of atoms containing unpaired electrons that reacts with oxygen and other reactive chemicals.

Functional disorder   A disease affecting a function of an organ, often without an apparent organic change.

Glycosides, cardiac   Organic chemicals containing an aglycone and a carbohydrate portion. Some cardiac glycosides stimulate contractibility of heart muscle, such as digitalis.

Hallucinogen   A drug that can induce mental hallucinations, and sometimes psychoses and insanity.

Hibernation, artificial   Cooling the body by immersion in cold water or by drugs.

Histamine   A neurohormone involved in allergies and some other disease conditions.

Hormone   A chemical formed in one organ of the body and carried by the circulation to another tissue, where it exerts a catalytic biological action. Many synthetic hormone analogues are known.

Hyperacidity   Excess gastric acid.

Hyperglycemia   *See* Diabetes.

Hypoglycemia   Abnormally low blood sugar.

Hypophysis   *See* Pituitary gland.

Immune system   Cellular biochemicals that protect the body against some diseases.

Involuntary organ   Glands, nerves, and muscles over which one has no control.

Isotonic   Having the same osmotic pressure and salt concentration as blood.

Leishmaniasis   Tropical protozoan infection caused by *Leishmania*.

Metastasis   Migration and relocation of malignant (cancer) cells.

Microbiology   Science of the chemistry and biology of microbes.

Molecular modification   Planned chemical variation of the structure of a prototype drug. Its aim is to improve therapeutic properties.

Mutation   A variation of inheritable characteristics caused by changes in the nucleotide sequence of genes.

Mycobacteria   Bacteria naturally encased in a capsule of lipids (fats). They cause slow infections such as leprosy and tuberculosis.

Mycoses   Fungus-caused diseases.

Mydriasis   Prolonged or excessive dilation of the pupil of the eye, caused by eye diseases or by a dilating drug.

Nerves, parasympathetic and sympathetic   Systems of autonomic nerves whose impulses are transmitted by neurotransmitters across the synapse.

Neuralgia   Superficial or surface pain.

Neurohormone   A hormonal chemical secreted at nerve endings and involved in the transmission of nervous impulses.

Neuroleptic   A tranquilizing drug; an antipsychotic.

Neurotransmitter   A chemical substance bridging conductance of electric potential along nerves.

Nonsteroidal antiinflammatory drug   A drug useful in arthritis and other inflammatory conditions, such as salicylates, indomethacin, ibuprofen, and sulindac.

Nutrients, essential   Substances in the diet without which deficiency symptoms appear.

Oncogene   Cancer inducer.

Opiates   Compounds derived from, or similar in action to, potent analgesic opium alkaloids. Many potent synthetic analgesics and euphorics are wrongly called opiates.

Parenteral drugs   Drugs administered by injection.

Parkinson's disease   Neurological disorder (shaking palsy) accompanied by dopamine deficiency. Patients exhibit extrapyramidal symptoms.

Pathogen   An agent or cell that causes disease.

Pathology   The branch of medicine that studies structural and functional changes caused by disease. It is often used to mean the disease process.

Penicillin   A large group of antibacterial antibiotics, some of which are semisynthetic chemicals. Chemically, all are beta-lactams.

Peptide   A combination of amino acids. A *dipeptide* is a combination of two amino acids; a *polypeptide*, of many amino acids.

Pharmacology   The study of the biochemistry, uses, and biological and therapeutic effects of drugs (not to be confused with pharmacy).

Pharmacopoeia   An official compendium listing medicinal drugs, their properties, standards of purity, and other useful information.

"Pill," the (contraceptive)   A mixture of estrogen and synthetic progestins that controls menstrual cycles and produces a state of pseudopregnancy, thereby preventing conception.

Pinworm   A scalp disease caused by *Oxyuris* worms.

Pituitary gland   A small endocrine (hormone-secreting) gland at the base of the brain, affecting metabolism, milk production, body growth, and behavior. The gland is also known as the hypophysis.

Placebo   A preparation containing no medication, but given for its psychological effect.

Prostaglandin   A chemically related family of lipid carboxylic acids with 20 carbon atoms, derived from arachidonic acid. They have multiple biological and therapeutic functions.

Psychosis   A form of mental illness. Also a symptom of schizophrenia.

Psychosomatic disease   Physical disorder of the body originating in or aggravated by psychic or emotional activity.

Psychotomimetic   A chemical that produces symptoms similar to pathological psychoses, such as amphetamine, LSD, and mescaline.

Purines, pyrimidines   Hexagonal, ring-shaped chemical structures found in nucleic acids.

Quats   Ammonium salts with antimicrobial properties. "Quats" is short for quaternary ammonium salts.

Recombinant DNA   A technique of inserting a foreign gene into the biosynthesis of a DNA, forcing the cell to make the foreign material. It is used in the manufacture of human pancreatic insulin by bacteria or yeasts.

Schistosomiasis   An infestation caused by schistosome worms that invade body organs.

Schizophrenia   A mental disease ("split personality") in which the patient exhibits withdrawal, delusions of grandeur and persecution, hallucinations, and sometimes psychoses.

Sedative   A drug that decreases excitement, irritation, and fear.

Sex hormones   Chemicals produced by sex organs or occasionally by some other tissues. They determine sexual quality, expression, and behavior and sexual development and processes. Estrogens are female sex hormones; androgens are male sex hormones.

Side effects   Unplanned effects of a drug.

Spirochete   Slender, spiral-shaped bacteria, some of which cause diseases (such as syphilis).

Substrate   A chemical that undergoes reactions, usually by being acted upon by an enzyme.

Suicide enzyme inhibitor   A chemical that reacts as both substrate and inhibitor of an enzyme, using up the enzyme and wasting its catalytic action.

Synapse   The minute gap between nerve dendrites across which nerve impulses are transmitted by neurohormones.

Teratogen   A chemical that causes birth defects.

Tetracyclines   A group of broad-spectrum antibiotics with a four-ring structure of atoms.

Thyroid   A ductless (endocrine) gland in front and on either side of the windpipe that regulates oxidative processes and growth of the body.

Toxicity   Degree of being poisonous.

Trademark   A registered symbol protected by law and used by a manufacturer to distinguish his or her product from those of competitors.

Trichomoniasis   An infectious disease caused by a protozoan microbe called *Trichomonas vaginalis*.

Trypanosome   Flagellate protozoans transmitted to humans and other vertebrates by insect vectors; they often cause disease.

Trypanosomiasis   Diseases caused by trypanosomes, e.g., sleeping sickness, Chagas' disease.

Twilight sleep   State of semiconsciousness induced by certain drugs, e.g., scopolamine. It can lessen pain.

Vasoconstrictor   A chemical that contracts (narrows) blood vessels.

Virus   Complex chemicals consisting of nucleic acids wrapped in a protein capsule of very high molecular weight. A virus can multiply like animate cells at the expense of its host cell, causing disease.

Vitamins   Several chemically different substances that regulate many developmental and bodily functions. Vitamins cannot be biosynthesized in the body and must be obtained from the diet or from synthetic diet supplements.

X-ray diffraction   A method of determining the chemical structure of a compound by measuring the distances between its atoms.

Index

Copy editing and production: Paula M. Bérard
Acquisition: Barbara E. Pralle

Production Manager: Cheryl Wurzbacher

Composition: Betsy Kulamer, Washington, DC
Indexing: Indexing Plus, Richmond, VA
Cover design: Marshall Henrichs, Lexington, MA
Printing and binding: Maple Press, York, PA